W0090536

THE MAMMALS OF
NORTH AMERICA

The Viviparous Quadrupeds of North America (its original title) was first published in three volumes between 1845 and 1848. There were 150 folio plates in total, each hand printed and coloured under the direction of J. T. Bowen of Philadelphia and measuring approximately 22 x 28 in (56 x 71 cm). The three volumes of plates were accompanied by three text volumes which were published between 1846 and 1854. The whole project was a collaboration between John James Audubon and the Reverend John Bachman, a leading American mammalogist. This edition has been created by photographing the volumes held by the Natural History Museum, London. The original numbering and captions have been faithfully recreated, and for this reason there are inconsistencies in the spelling of names and numbering of plates. The list of current common and species names at the back of the book has been compiled by Professor Kristofer M. Helgen, The University of Adelaide and Roberto Portela Miguez, Senior Curator Mammals, Natural History Museum, London with reference to Mammal Species of the World: a Taxonomic and Geographic Reference by D. E. Wilson and D. M. Reeder (Johns Hopkins University Press, Baltimore, 2005) and more contemporary publications for individual cases.

First published by the Natural History Museum,
Cromwell Road, London SW7 5BD
© The Trustees of the Natural History Museum, London 2019
Foreword © Kristofer Helgen, 2019

ISBN 978 0 565 09484 3

All rights reserved. No part of this publication may be transmitted in any form
or by any means without prior permission from the British Publisher.

A catalogue record for this book is available from the British Library.

REPRODUCTION BY DL IMAGING
PRINTED BY 1010 PRINTING INTERNATIONAL LIMITED

End paper: © Tom Grundy/Shutterstock.com

JOHN JAMES AUDUBON
THE MAMMALS OF NORTH AMERICA

Published by the Natural History Museum, London

Foreword

JOHN JAMES AUDUBON is one of the most famous names in natural history. Born in modern-day Haiti in 1785, Audubon was raised in France and spent most of his life in the fledgling United States, travelling widely and living in Pennsylvania, Kentucky, Louisiana, New York, and elsewhere. He was a naturalist who spent a lifetime relentlessly seeking to uncover the lifeways of wildlife, especially American birds and mammals. He was an American frontiersman who enjoyed a long and loving marriage to his wife, Lucy, with whom he raised two sons, Victor and John Woodhouse. Both sons, like their father, would become naturalists and artists. Audubon's masterwork, *The Birds of America*, a beautiful union of art and science, is one of the most famous books ever published. He travelled extensively in North America and in Europe, and his last expedition, in 1843, took him as far west as the Yellowstone River. He died in 1851, at the age of 65.

Not everything for Audubon was a success, and little in his life was straightforward. He struggled for many years to achieve his obsessive goal of publishing *The Birds of America*. He was an unfaithful husband, a difficult business partner, and he failed with many financial ventures, even serving time in a debtor's prison. He knew tragedy; both of his daughters, Lucy and Rose, died in infancy. And he was notoriously loose with the truth. Audubon was the illegitimate son of a French ship captain and a Creole chambermaid, and his mother died soon after his birth. He always sought to hide the truth of his origins, which were embarrassing to him, claiming he was born in Louisiana and that his mother was a 'lady of Spanish extraction, as beautiful as she was wealthy.' He falsely claimed that he studied under the great French painter Jacques Louis David, that he shared adventures with the frontiersman Daniel Boone, and that his father fought alongside General George Washington. Part of Audubon's allure, then and now, is the mythology attached to him, some of which was of his own creation. He used embellishments and fabrications to burnish his larger-than-life credentials on both sides of the Atlantic. Frontiers, after all, are the province of tall tales.

In the aftermath of *The Birds of America* and its success, Audubon aspired to produce a similar publication featuring paintings and accounts of the mammals of North America. For this he would need help. One of Audubon's friends was the Reverend John Bachman, a leading American mammalogist, who presided over St. John's Lutheran Church in Charleston, South Carolina. Bachman was a gifted amateur natural historian and taxonomist, responsible for original scientific names first provided to many of the smaller mammals of North America, including moles, shrews, squirrels, voles, rabbits, and hares. Audubon was a seasoned observer of mammals but not a rigorous taxonomist; most of the mammals Audubon named were squirrels that later proved to be colour variants of previously described species – variations in colour that allowed ornithologists to distinguish bird species could be misleading in mammals. (The Reverend humoured Audubon's opinions on squirrels, which he co-authored, but joked of divine difficulties, warning Audubon that 'the ever-varying squirrels were sent by the old boy himself, to puzzle the naturalist.') Audubon would need Bachman's expertise, and invited him to collaborate to produce his planned book on mammals.[1]

Audubon and Bachman chose to title their work *Viviparous Quadrupeds of North America*. Not simply *Mammals*, because they ultimately decided not to include bats and marine mammals (which, technically speaking, were not creditable as *quadrupeds*). Nor simply *Quadrupeds*, because it would not include reptiles and amphibians (which have four legs but, laying eggs, are *oviparous*). In later editions the title shed its unwieldy first name, becoming *Quadrupeds of North America*. The publication was a collaboration not just between the two men themselves, but between the intertwined Audubon and Bachman families. Audubon's two sons married two of Bachman's daughters – John Woodhouse wed Maria Bachman in 1837, and Victor married Eliza Bachman in 1839. Both marriages were sadly short lived, as Maria died in 1840 and Eliza died in 1841, their young lives both cut short by tuberculosis. As Audubon's own health and artistic command failed with age, Victor and John Woodhouse took over the job of painting from their father – Victor painting mostly background scenes, and John painting almost half of the mammal portraits that appeared in *Quadrupeds*. The Audubons thus provided the artwork, and Bachman wrote most of the text that accompanied the paintings, drawing in part on observations from the Audubons' journals.

Before his abilities faltered, many of Audubon's portraits of mammals were as masterful as those of his birds, elegant captures of poise and motion in an age before photography. Some, like his American badger and his Swift fox, were drawn from life, from animals he kept as pets. Others, mostly species from the eastern United States, were drawn from museum specimens, supplemented ably by Audubon's decades-long experience with them as wild animals. Some leap to life from the page and recall particular places and moments. Audubon's firsthand observations of the 'American Bison or Buffalo', published in *Quadrupeds*, were made a generation before the wholesale slaughter of the bison emptied the prairies of their thundering abundance, and his paintings of bison on the prairie capture scenes of a world that would soon be forever transformed. Similarly, his painting of the wolverine poignantly depicts an emblem of American northern wilderness. Audubon's portrait shows a half-grown animal in a combative pose, with body tensed, mid-snarl, as a hunter might encounter it. Even in the early nineteenth century, the wolverine's distribution had shrunk from well-settled areas of North America, and it was largely gone from the northeastern part of the United States. Much of Audubon's characteristic insight regarding mammals appears as stories of hunting and shooting; yet his fondness and empathy for animals often shines through. Describing a captive wolverine observed in Europe, Audubon and Bachman wrote 'He was very gentle, opened his mouth to let us examine his teeth, and buried his head in our lap as we admired his long claws and felt his woolly feet.' In this endearing scene we are far removed from the portrait's snarls.

While never rising to the supreme standing of Audubon's *Birds*, *Quadrupeds of North America* is a great work of American natural history. An ornithologist to his core, Audubon knew much less about mammals than birds, and the limitations on his firsthand observations (most mammals are nocturnal and small, often harder to observe, distinguish, and illustrate than birds) and his command of mammalian behaviour and anatomy render the paintings more earthbound than his soaring *Birds*. Further, relatively few original scientific discoveries and observations about mammals ultimately appeared in *Quadrupeds*.[2] Some subscribers to *Quadrupeds* groused over the seeming overabundance of squirrel varieties, a weakness, as noted above, that has not stood up to modern taxonomic accounting. But much of the challenges involved were simply a matter of timing. The mammals of eastern North America were reasonably well known to science at the time, and could be admirably painted and reviewed by Audubon and Bachman, but those of the west and southern parts of North America were only just coming into view.[3] Audubon's advancing age also meant that he could paint only half of the portraits that appeared in *Quadrupeds*, and this half included mostly small species, like rodents and shrews, and relatively few of the larger mammals, like deer, antelope, bears, and wolves, that most readers wanted to admire. His son John's portraits, which fill half of *Quadrupeds*, are rather clumsy in comparison, unfortunately lacking his father's artistic genius. Many are stilted, out of proportion, and have a taxidermic staleness, but others, like his armadillo and ocelot, are beautifully executed. Much of the scientific text in *Quadrupeds* also shows a debt to this younger Audubon's own mammalogical observations recorded in his journals from his own expedition to Texas, and to his firsthand study of mammal specimens in museums in Europe. Audubon's sons both became mammalogists themselves, and their devotion to their father ensured that they helped see his *Quadrupeds* through to completion.

I return regularly to the *Quadrupeds of North America*, both for its science and for its art. The paintings remind us that Audubon was a mammalogist as well as an ornithologist, and serve as a monument to one of the most famous partnerships in American natural history, that between Audubon and Bachman.[4] They remind us especially of the splendour of the varied wildernesses of North America – since Audubon's time drastically changed, but not yet lost.

Professor Kristofer M. Helgen, The University of Adelaide

[1] Today, both men each have a single recognized species of mammal named in their honour, both of which are cottontail rabbits – the Desert Cottontail, *Sylvilagus audubonii* (Baird, 1857), and the Brush Rabbit, *Sylvilagus bachmani* (Waterhouse, 1839), both species of western North America.

[2] There were a few remarkable discoveries included, however, the most important being the first scientific description of the Black-footed ferret now classified as *Mustela nigripes* (Audubon and Bachman, 1851). Already rare in Audubon's time, the species is now extinct in the wild – a decline, like that of the American bison, is symptomatic of the broader environmental devastation of North America's vast western prairies. In *Quadrupeds*, Audubon and Bachman also introduced the scientific name *Canis rufus*, still widely used for the 'Red wolf' of North America, an animal also extinct in the wild today.

[3] In particular, *Quadrupeds* was quickly eclipsed as a scientific reference by more scholarly and comprehensive treatments of the North American mammal fauna, starting with Spencer Fullerton Baird's *Mammals of North America* in 1859, which had as its basis new federally funded surveys of the American West, and the systematic study of the scientific collections of the Smithsonian Institution – the young but burgeoning national museum of the United States.

[4] Further reading:
Rhodes, Richard, 2004. *John James Audubon: The Making of an American*. Knopf, New York.
Boehme, Sarah E. (editor), 2000. *John James Audubon in the West: The Last Expedition; Mammals of North America*. Abrams, New York.

PLATE 1.

Lith. Printed & Colᵈ by J.T. Bowen, Philᵃ 1841.

LYNX RUFUS, GÜLDENSTÆDT.
COMMON AMERICAN WILD CAT.
Natural Size
MALE.

Drawn from Nature by J.J. Audubon, F.R.S. F.L.S.

PLATE. II.

ARCTOMYS MONAX. GMEL.
MARYLAND, MARMOT, WOODCHUCK, GROUNDHOG.
Natural Size —
MALE & FEMALE

Drawn from Nature by J.J.Audubon F.R.S.F.L.S.

Lith. Printed & Col.d by J.T. Bowen, Phila. 1842.

N.o 1.

PLATE. III.

LEPUS TOWNSENDII BACH.

TOWNSEND'S ROCKY MOUNTAIN HARE.

Natural size.

MALE & FEMALE.

Drawn from Nature by J.J.Audubon,F.R.S.F.L.S.

Lith. Printed & Col.ᵈ by J.T.Bowen, Phila. 1842.

NEOTOMA FLORIDANA. SAY ET ORD.

FLORIDA RAT

Natural Size

MALE, FEMALE AND YOUNG OF DIFFERENT AGES.

PLATE. V.

Drawn from Nature by J.J. Audubon F.R.S. F.L.S.

Lith. Printed & Col.d by J.T. Bowen, Phila. 1842.

SCIURUS RICHARDSONII, BACH.

RICHARDSON'S COLUMBIAN SQUIREL.

Natural Size

MALE & FEMALE.

PLATE VI.

Drawn from Nature by J. J. Audubon, F. R. S. F. L. S.

CANIS (VULPES) FULVUS. DESMAREST.

VAR. DECUSSATUS.

AMERICAN CROSS FOX.

Natural Size.

MALE.

Lith. Printed & Col.d by J. T. Bowen, Philad.a 1845.

SCIURUS CAROLINENSIS. GMELIN.

CAROLINA GREY SQUIRREL.

Natural Size.

MALE AND FEMALE.

Drawn from Nature by J.J.Audubon, F.R.S.F.L.S.

Lith. Printed & Col.d by J.T. Bowen, Phila. 1843.

PLATE VIII.

Drawn from Nature by J.J. Audubon. F.R.S. F.L.S.

Lith. Printed & Col.ᵈ by J.T. Bowen. Philad.ᵃ

TAMIAS LYSTERI. RAY.
CHIPPING SQUIRREL. HACKEE &c.
Natural Size.
MALE, FEMALE AND YOUNG FIRST AUTUMN.

No. 2.

SPERMOPHILUS PARRYI, RICHARDSON.

PARRY'S MARMOT SQUIRREL.

Natural Size.

Drawn from Nature by J.J.Audubon, F.R.S.L.S.

Lith.Printed &col.d by J.T.Bowen, Phila.d 1843.

PLATE X.

SCALLOPS AQUATICUS, LINN.

COMMON AMERICAN SHREW MOLE

Natural Size.

MALE AND FEMALE.

Drawn from Nature by J.J. Audubon, F.R.S.F.L.S.

Lith. Printed & Col.d by J.T. Bowen, Phila. 1843.

PLATE XI.

Lith. Printed & Col.d by J.T. Bowen, Phila. 1848.

LEPUS AMERICANUS, ERXLEBEIN.
NORTHERN HARE. Summer.
Natural Size
L. MALE, 2. FEMALE.

Drawn from Nature by J.J. Audubon, F.R.S.F.L.S.

Nº 3.

PLATE XII.

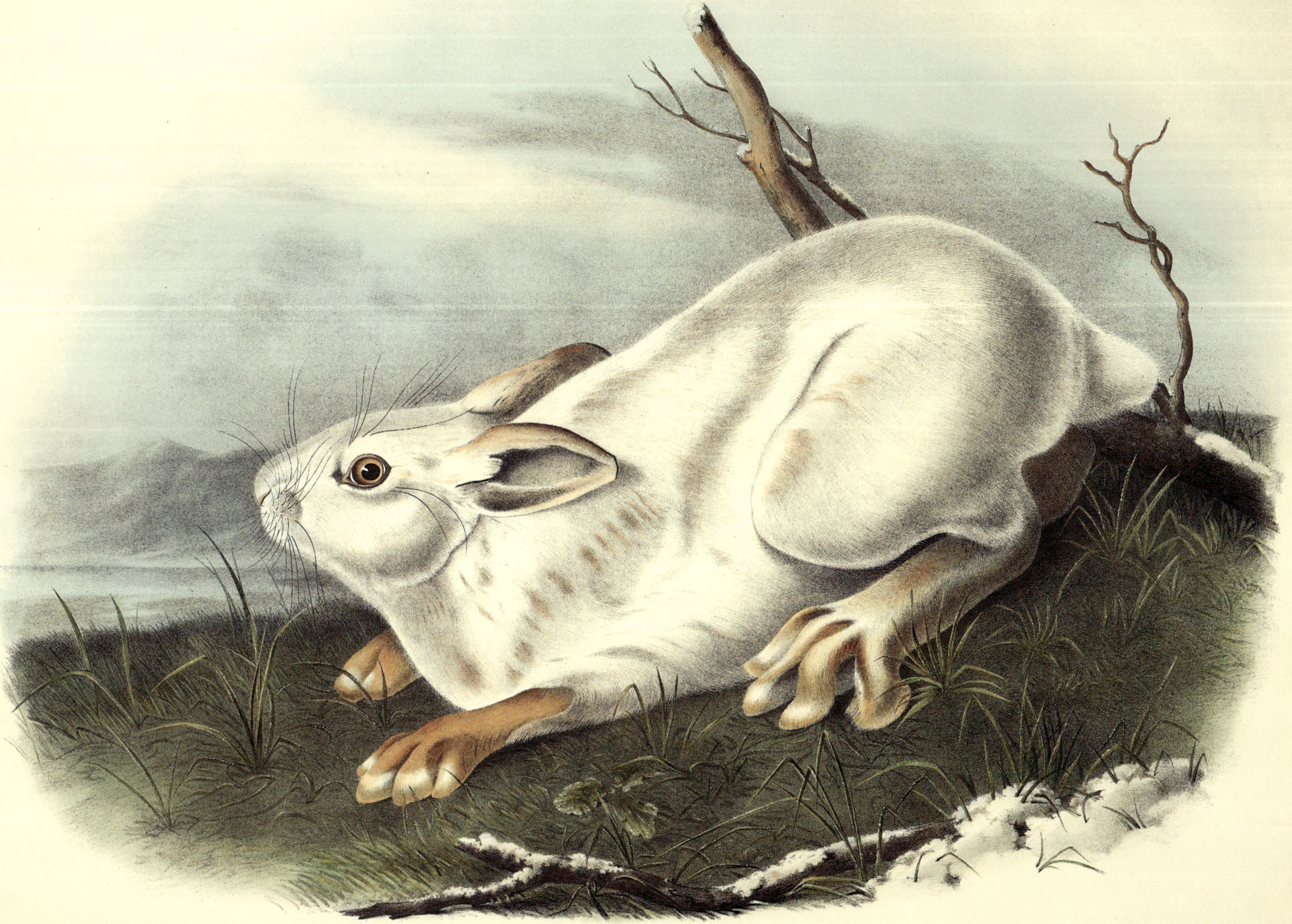

Drawn from Nature by J. J. Audubon, F.R.S. F.L.S.

Lith. Printed & Col.d by J. T. Bowen, Phila. 1843.

LEPUS AMERICANUS, ERXLEBEIN.

NORTHERN HARE.

Natural Size

WINTER.

PLATE XIII.

FIBER ZIBETHICUS, CUVIER.

MUSK-RAT, MUSQUASH.

Natural Size.

OLD AND YOUNG.

Drawn from Nature by J.J. Audubon, F.R.S. F.L.S.

Lith. Printed & Col.d by J.T. Bowen, Phila. 1845.

Drawn from Nature by J.J.Audubon, F.R.S.F.L.S.

Lith. Printed & Col.d by J.T.Bowen, 1843.

SCIURUS HUDSONIUS, PENNANT.

HUDSON'S BAY SQUIRREL CHICKAREE RED SQUIRREL.

Natural Size

PLATE XV.

Drawn from Nature by J. J. Audubon, F.R.S.F.L.S.

Lith Printed & Col.d J.T.Bowen, Phila.

PTEROMYS OREGONENSIS, BACHMAN.

OREGON FLYING SQUIRREL.

Natural Size.

PLATE XVI.

No 4.

LYNX CANADENSIS, GEOFF.

CANADA LYNX.

Nature &c.

Drawn from Nature by J.J.Audubon,F.R.S.F.L.S.

Lith.Printed & Col.d by J.T.Bowen.Philad.a

Drawn from Nature by J.J. Audubon, F.R.S.F.L.S.

Lith. Printed & Col.d by J.T. Bowen, Phila. 1843.

Sciurus cinereus. Linn. Gmel.

CAT SQUIRREL

Natural Size

PLATE XVIII.

Lith. Printed & Col.ᵈ by J.T. Bowen, Phila. 1842.

Drawn from Nature by J.J. Audubon, F.R.S. F.L.S.

LEPUS PALUSTRIS, BACHMAN.

MARSH HARE.

Natural Size.

Drawn from Nature by J.J. Audubon, F.R.S.F.L.S.

Lith. Printed & Col⁴ by J.T. Bowen, Phila.

SCIURUS MOLLIPILOSUS, AUD. AND BACH.
SOFT-HAIRED SQUIRREL.
Natural Size.

PLATE. XX.

Drawn from Nature by J.J.Audubon, F.R.S.F.L.S.

Lith.Printed & Col.d by J.T.Bowen, Phila.1843.

TAMIAS TOWNSENDII, BACHMAN.

TOWNSEND'S GROUND SQUIRREL.

Natural Size.

No. 5.

PLATE XXI.

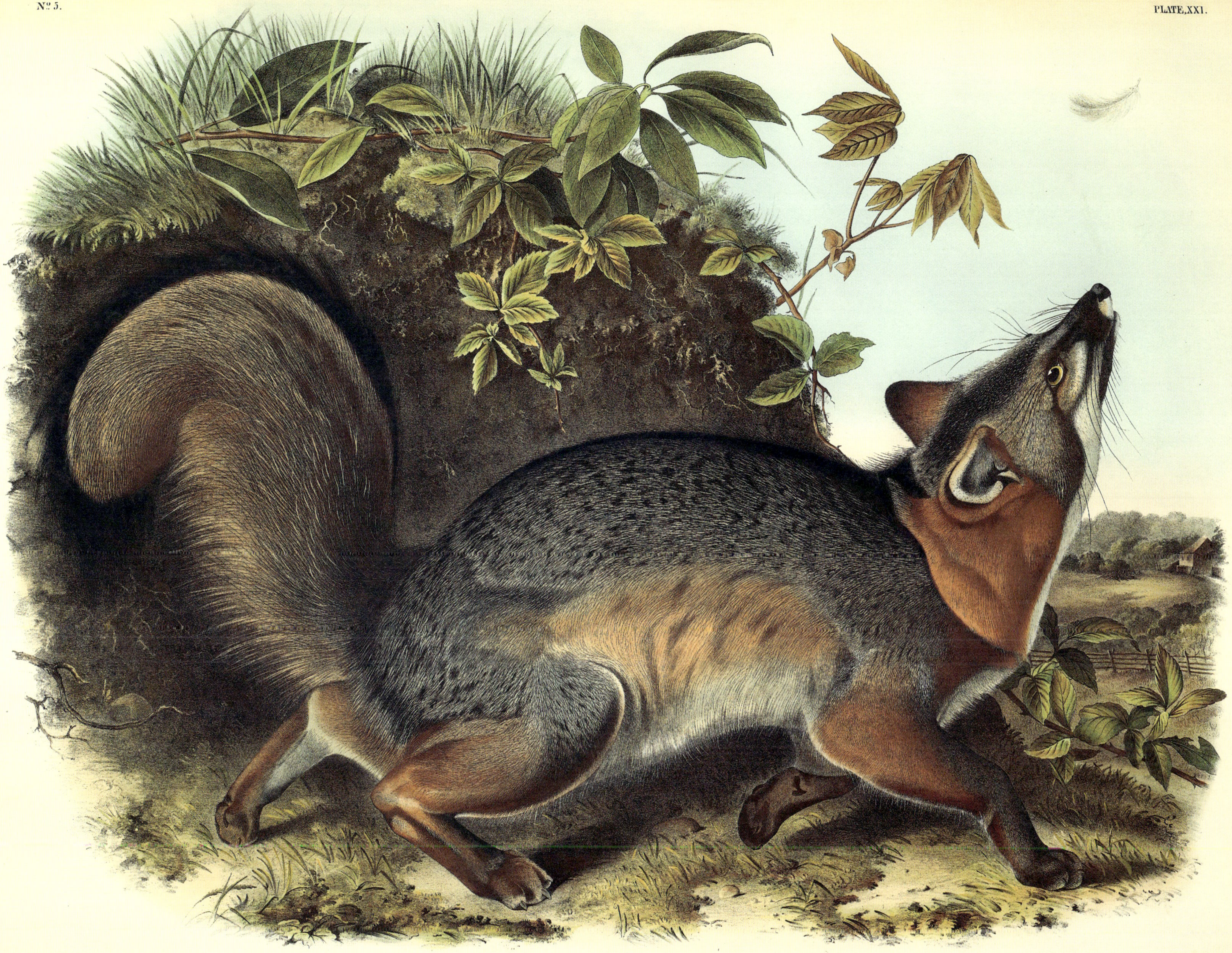

Drawn from Nature by J. J. Audubon, F.R.S. F.L.S.

CANIS (VULPES) VIRGINIANUS, GMEL.

GREY FOX

5/7 Natural Size

Lith. Printed & Col.d by J. T. Bowen, Philad.a 1843.

PLATE. XXII.

LEPUS SYLVATICUS, BACHMAN
GREY RABBIT.
Natural Size.
OLD & YOUNG.

Drawn from Nature by J. J. Audubon, F.R.S. F.L.S.

Lith. Printed & Col'd by J. T. Bowen, Philad.ª 1842.

PLATE XXIII.

MUS RATTUS ET VAR. LINN.

BLACK RAT.

Natural size.

OLD & YOUNG.

Drawn from Nature by J.J.Audubon, F.R.S. F.L.S.

Lith Printed & Col.d by J.T.Bowen, Phila. 1843.

PLATE XXIV.

TAMIAS QUADRIVITTATUS, SAY.

FOUR-STRIPED GROUND SQUIRREL.

Natural Size.

1. MALE. 2. FEMALE. 3. 4. 5. YOUNG.

Drawn from Nature by J.J. Audubon, F.R.S.F.L.S.

Lith. Printed & Col. by J.T. Bowen, Philad.ª 1843.

PLATE. XXV.

Drawn from Nature by J.J.Audubon.F.R.S.FL.S.

Lith. Printed & Colᵈ by J.T.Bowen, Philadᵃ 1843.

SCIURUS LANIGINOSUS, BACH.

DOWNY SQUIRREL.

Natural Size.

PLATE XXVI.

GULO LUSCUS, LIN.

THE WOLVERINE.

Natural size.

Drawn from Nature by J. J. Audubon, F.R.S.F.L.S.

Lith. Printed & Col.d by J. T. Bowen, Philad.a 1845.

PLATE. XXVII.

Drawn from Nature by J. J. Audubon, F.R.S. F.L.S.

Lith. Printed & Col.ᵈ by J. T. Bowen, Philad.ᵃ 1844.

SCIURUS LONGIPILIS, AUD & BACH.

LONG HAIRED SQUIRREL.

Natural Size.

Drawn from Nature by J. J. Audubon, F.R.S.F.L.S.

Lith. Printed & Col.d by J. T. Bowen, Philad.a 1843.

PTEROMYS VOLUCELLA, GMEL.
COMMON FLYING SQUIRREL.
Natural Size.
1. 2. MALES. 3. 4. FEMALES. 5. YOUNG.

PLATE XXIX.

Drawn from Nature by J.J.Audubon, F.R.S. F.L.S.

Lith. Printed & Col.d by J.T Bowen, Philad.a 1843.

NEOTOMA DRUMMONDII, RICH.

ROCKY MOUNTAIN NEOTOMA.

Natural Size.

PLATE. XXX.

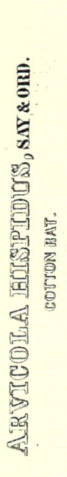

Drawn from Nature by J. J. Audubon, F.R.S. F.L.S.

ARVICOLA HISPIDUS, SAY & ORD.

COTTON RAT.

Natural Size.

Lith. Printed & Col.d by J. T. Bowen, Phila.ᵈ 1843.

PLATE XXXI.

DICOTYLES TORQUATUS, F. CUV.

COLLARED PECCARY.

Natural Size.

Drawn from Nature by J. J. Audubon, F.R.S.L.

Lith. Printed & Col'd by J. T. Bowen, Philad.ª 1844.

No 7. PLATE XXXII.

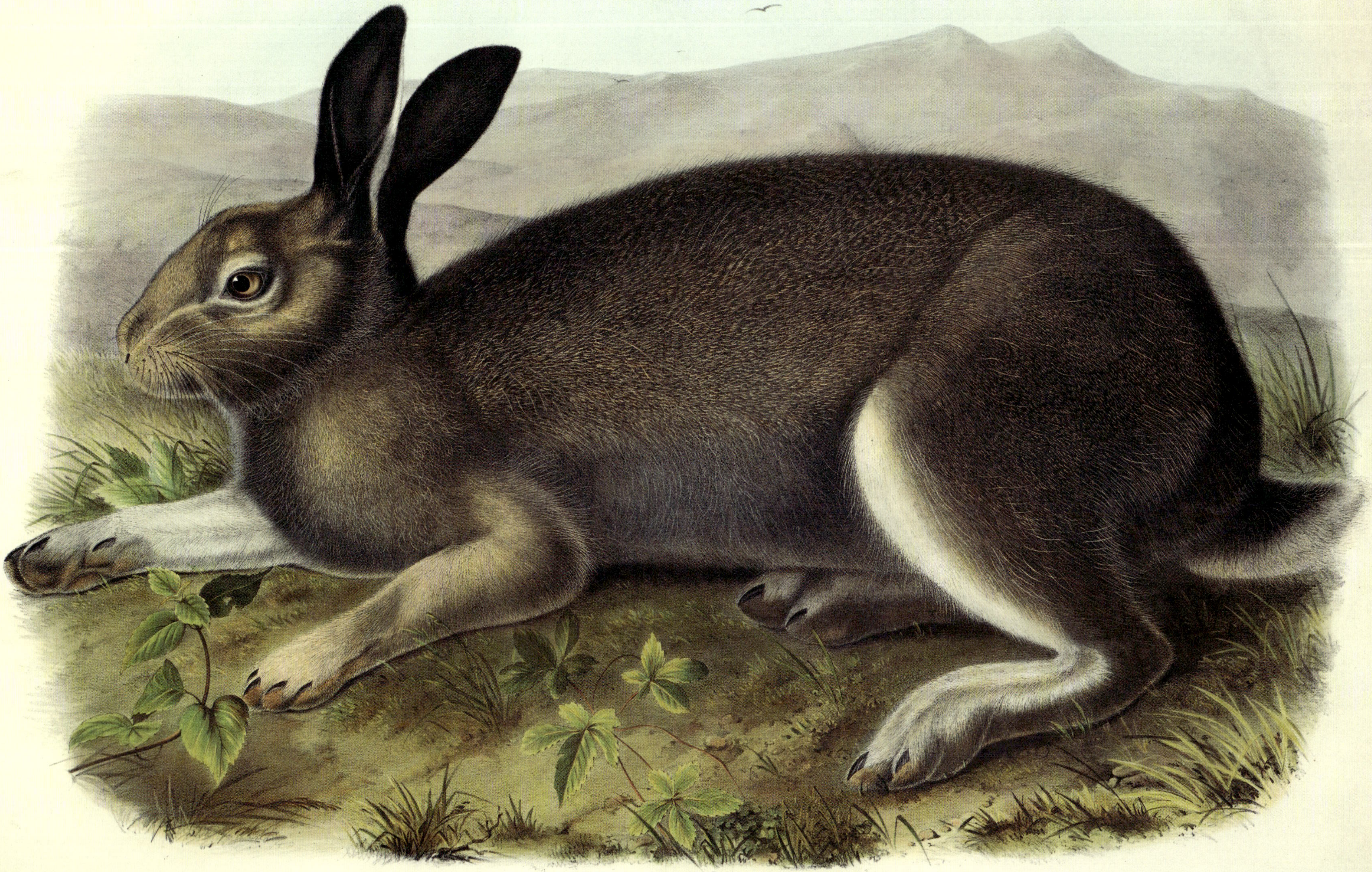

Drawn from Nature by J. J. Audubon, F.R.S. F.L.S.

Lith. Printed & Col.d J. T. Bowen, Philad.a 1844.

LEPUS GLACIALIS, LEACH.

POLAR HARE.

Natural Size.

PLATE XXXIII.

Drawn from Nature by J.J. Audubon, F.R.S. F.L.S.

Lith. Printed & Col.d by J.T. Bowen, Philad. 1844.

PUTORIUS VISON, LINN.

MINK.

Natural Size.

MALE & FEMALE.

Drawn from Nature by J. J. Audubon, F. R. S. F. L. S.

Lith. Printed & Col.ᵈ by J. T. Bowen, Philad.ᵃ 1844.

SCIURUS NIGER, LIN.

BLACK SQUIRREL.

Natural Size.

1. MALE. 2. FEMALE.

SCIURUS MIGRATORIUS, BACH.

MIGRATORY SQUIRREL.

Natural Size

1. OLD MALE. 2. FEMALE. 3. YOUNG.

Drawn from Nature by J. J. Audubon, F.R.S. F.L.S.

Lith. Printed & Col.d by J. T. Bowen, Philad.a 1844.

Drawn from Nature by J.J.Audubon, F.R.S.F.L.S.

Lith. Printed & Col.d by J. T. Bowen, Philad.a 1844.

HYSTRIX DORSATA, LINN.
CANADA PORCUPINE.
½ Natural Size.

PLATE XXXVII.

LEPUS AQUATICUS, BACH.

SWAMP HARE.

Drawn from Nature by J.J. Audubon, F.R.S.F.L.S.

Lith. Printed & Col'd by T. Bowen, Phila. 1844.

PLATE. XXXVIII.

Drawn from Nature by J.J.Audubon. F.R.S.F.L.S.

Lith. Printed & Col.d by J.T.Bowen.Phila.1844.

SCIURUS FERRUGINIVENTRIS, AUD. AND BACH.

RED-BELLIED SQUIRREL.

Natural Size.

MAN, FEMALE AND YOUNG.

PLATE XXXIX.

Drawn from Nature by J. J. Audubon, F.R.S. F.L.S.

Lith. Printed & Cold by J. T. Bowen, Philad.ª 1844.

SPERMOPHILUS TRIDECEMLINEATUS, MITCHELL.

Leopard Spermophile.

Natural Size.

MALE AND FEMALE.

PLATE XI.

Drawn from Nature by J.J. Audubon, F.R.S. F.L.S.

Lith. Printed & Col.ᵈ by J.T. Bowen, Phila.ᵈ 1844.

MUS LEUCOPUS, RAFF.
WHITE FOOTED MOUSE.
Natural Size.
MALE, FEMALE AND YOUNG.

MUSTELA CANADENSIS, LINN.

PENNANT'S MARTEN OR FISHER.

Natural Size.

MALE.

Drawn from Nature by J.J Audubon F.R.S.F.L.S

Lith.ᵈ Printed & Col.ᵈ by J.T. Bowen Philad.ᵃ 1844

PLATE XLII.

Drawn from Nature by J. J. Audubon, F.R.S. F.L.S.

Lith. Printed & Col.d by J. T. Bowen, Philad.a 1844.

MEPHITIS AMERICANA, DESM.
COMMON AMERICAN SKUNK
Natural Size
FEMALE.

PLATE. XLIII.

Drawn from Nature by J. J. Audubon. F.R.S. F.L.S.

Lith. Printed & Col.d by J. T. Bowen, Philad.a 1844.

SCIURUS LEPORINUS, AUD. & BACH.
HARE SQUIRREL.
Natural Size.

PLATE XLIV.

PSEUDOSTOMA BURSARIUS, SHAW.

CANADA POUCHED RAT.

Drawn from Nature by J. J. Audubon, F.R.S. F.L.S.

Lith. Printed & Col'd by J. T. Bowen, Phila. 1844.

PLATE XLIV.

ARVICOLA PENNSYLVANICUS, ORD.

WILSON'S MEADOW MOUSE

Natural Size

Drawn from Nature by J. J. Audubon, F.R.S. F.L.S.

Lith. Printed & Col.d by J. T. Bowen, Philad.a 1844.

N°10. PLATE XLVI.

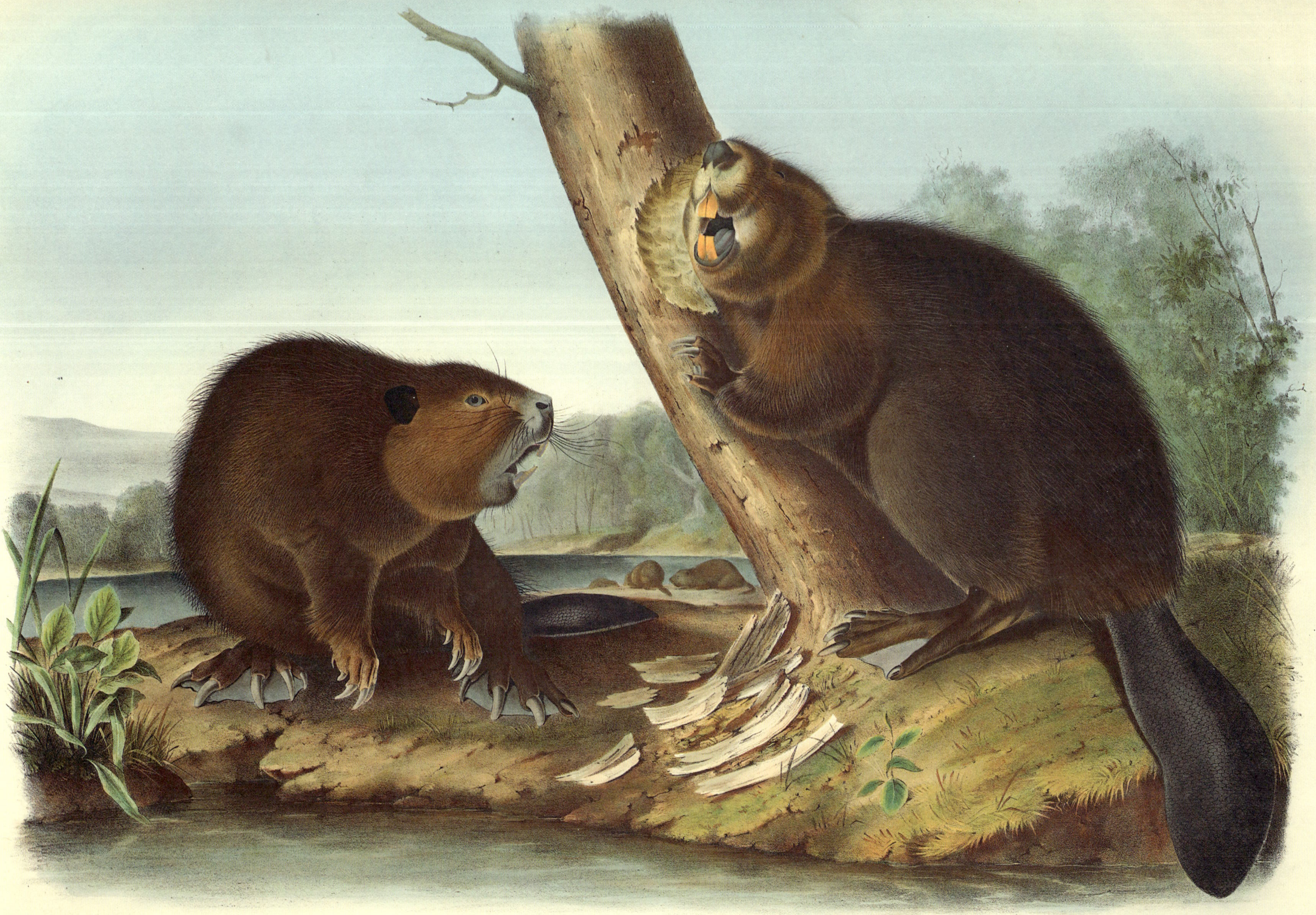

Drawn from Nature by J. J. Audubon, F.R.S. F.L.S. Lith. Printed & Col.d by J. T. Bowen, Philad.a 1844.

CASTOR FIBER AMERICANUS, LINN.

AMERICAN BEAVER.

2/3 Natural Size

PLATE XLVII.

MELES LABRADORIA, SABINE

AMERICAN BADGER.

Natural Size

Drawn from Nature by J. J. Audubon, F.R.S. F.L.S.

Lith. Printed & Cold by J. T. Bowen, Philada 1844.

PLATE. XLVIII.

Drawn from Nature by J.J. Audubon, F.R.S. F.L.S.

Lith. Printed & Col.d by J.T. Bowen, Phila. 1844.

SCIURUS DOUGLASSII, GRAY.

DOUGLASS SQUIRREL.

Natural Size.

1. MALE. 2. FEMALE.

PLATE XLIX

Drawn from Nature by J.J.Audubon, F.R.S.F.L.S.

Lith.ª Printed & Col.ª by J.T.Bowen, Philad.ª 1844

SPERMOPHILUS DOUGLASSII. RICHARDSON.

DOUGLASS'S SPERMOPHILE.

Natural Size.

PLATE L.

SPERMOPHILUS RICHARDSONII, SABINE.

RICHARDSON'S SPERMOPHILE.

Natural Size.

Drawn from Nature by J.J.Audubon. PLATE 5.

Lith. Printed & Col.d by J.T.Bowen, Phila.d 1844

PLATE LI.

Drawn from Nature by J.J.Audubon FRS&c.

Lith Printed & Cold by J.T.Bowen Philada 1844

LUTRA CANADENSIS, SABINE.

CANADA OTTER.

Natural Size.
MALE.

No. 11.

PLATE LII.

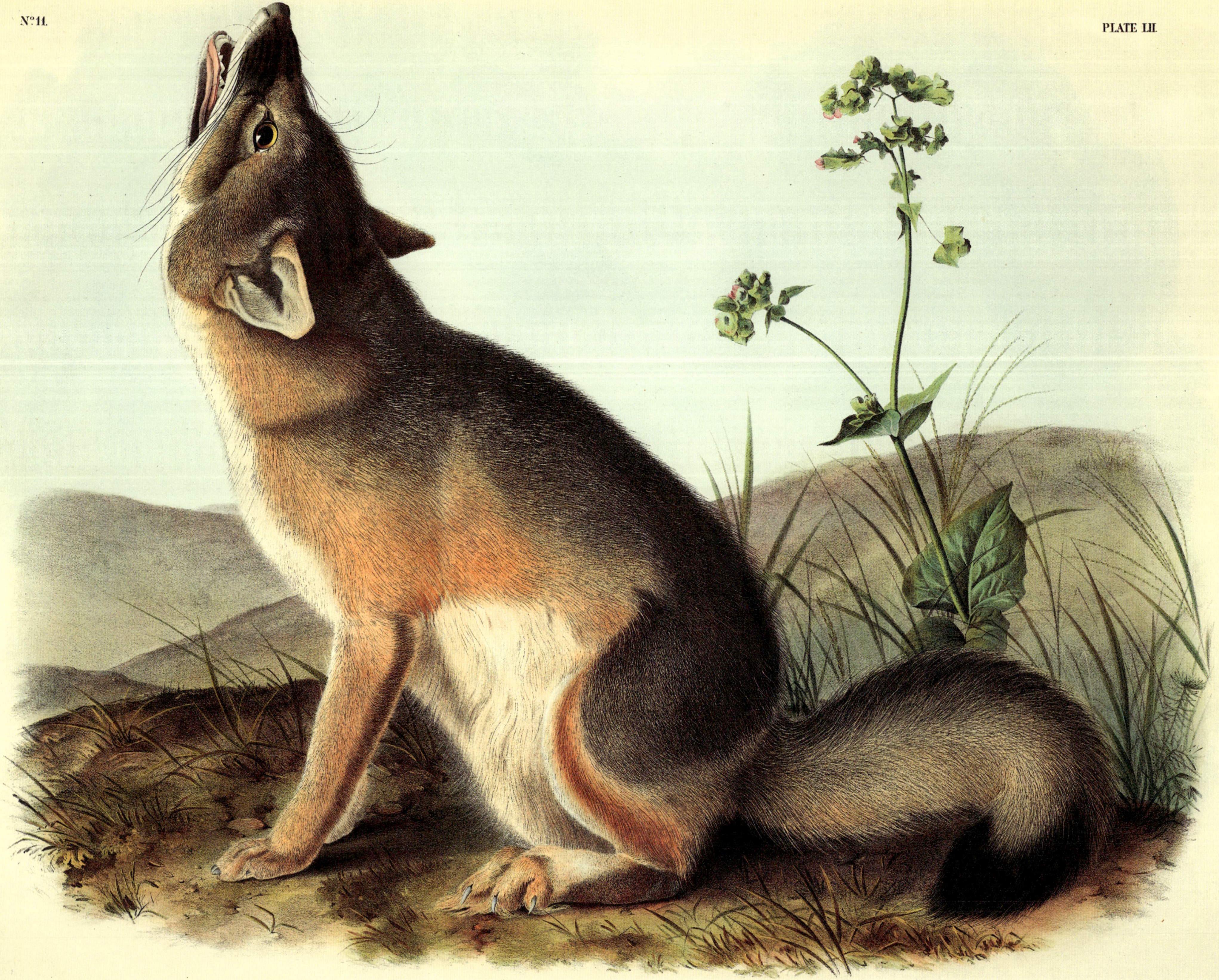

Drawn from Nature by J. J. Audubon, F.R.S.F.L.S.

VULPES VELOX, SAY.

SWIFT FOX.

Natural Size

MALE.

Lith. Printed & Col.d by J. T. Bowen, Philad.a 1844

PLATE LIII.

MEPHITIS MESOLEUCA, LICHT.

TEXAN SKUNK.

Natural Size.

Drawn from Nature by J.J.Audubon,F.R.S.F.L.S.

Lith.ᵈ Printed & Col.ᵈ by J.T.Bowen, Philad.ᵃ 1845.

PLATE LIV.

Lith. Printed & Col. by J.T. Bowen, Philad.1845.

MUS DECUMANUS, LINN

BROWN, OR NORWAY RAT.

MALE, FEMALE & YOUNG.

Natural Size.

Drawn from Nature by J.J.Audubon F.R.S.E.S.

Drawn from Nature by J.J.Audubon, F.R.S.F.L.S.

Lith. Printed & Col. by J.T. Bowen, Philad.ª 1844

SCIURUS RUBRICAUDATUS, AUD: & BACH:

RED-TAILED SQUIRREL.

Natural Size.

MALE.

PLATE LVI.

BOS AMERICANUS, GMEL.

AMERICAN BISON OR BUFFALO.

¼ Natural Size.

MALE.

Drawn from Nature by J.J. Audubon F.R.S.L.S.

Lith.d Printed & Col.d by J.T. Bowen, Phila.d 1845.

PLATE LVII.

Nº 12.

BOS AMERICANUS, GMEL.

AMERICAN BISON OR BUFFALO.

1. FEMALE, 2. YOUNG 3. MALE.

Drawn from Nature by J.J.Audubon F.R.S.F.L.S.

Lith. Printed & Col.by J.T.Bowen Philad.1845

Drawn from Nature by J.J.Audubon F.R.S.F.L.S.

Lith Printed & Col.^d by J. T. Bowen, Philad.^a 1846.

SCIURUS SUB-AURATUS, AUD: & BACH:

ORANGE-BELLIED SQUIRREL.

MALE & FEMALE.

Natural Size.

PLATE LIX.

Drawn from Nature by J. J. Audubon FRS. FLS.

PUTORIUS ERMINEA, LINN.

WHITE WEASEL. STOAT.

Natural Size.

1. MALE. 2. FEMALE.

Lith Printed & Col. by J. T. Bowen, Philad. 1845.

PLATE LX.

N°. 12.

Lith. Printed & Col. by J.T. Bowen, Philad. 1843.

Drawn from Nature by J.J. Audubon, FRS.E.L.

PUTORIUS FRENATA, LICHT.
BRIDLED WEASEL.
MALES.
Natural Size.

PLATE LXI

PROCYON LOTOR, CUVIER
RACCOON.
MALE.
Natural Size —

PLATE LXII.

No 13.

CERVUS CANADENSIS, RAY.

AMERICAN ELK = WAPITI DEER.

⅓ Natural Size

MALE AND FEMALE

Drawn from Nature by J.J.Audubon #237LS

Lith.Printed & Col.d by J.T.Bowen, Philad.1845

PLATE LXIII.

LEPUS NIGRICAUDATUS, BENNET.

BLACK-TAILED HARE.

MALE.

Natural Size.

Drawn from Nature by J.J.Audubon,F.R.S.F.L.S.

Lith.Printed & Col.d by J. T. Bowen, Phila.d 1845.

PLATE LXIV.

Drawn from Nature by J.J.Audubon, F.R.S.F.L.S.

Lith Printed & Col.d by J.T. Bowen, Philad.a 1845.

MUSTELA FUSCA, AUD. & BACH.

LITTLE AMERICAN BROWN WEASEL.

Natural Size.

Drawn from Nature by J.J.Audubon. F.R.S.F.L.S

MUS MINIMUS, AUD & BACH
LITTLE HARVEST MOUSE.
MALES & FEMALES.
Natural Size.

Lith.d Printed & Col.d by J.T.Bowen Philad.d 1845.

PLATE LXVI.

Lith. Printed & Col.d by J.T.Bowen, Phila.s 1845.

No. 14.

DIDELPHIS VIRGINIANA, PENNANT.
VIRGINIAN OPOSSUM.
FEMALE & YOUNG HALF, 7 MONTHS OLD.
Natural Size.

Drawn & Noted by J. J. Audubon, F.R.S.F.L.S.

PLATE LXVII.

Drawn from Nature by J.J.Audubon. F.R.S.F.L.S.

Lith. Printed & Col.d by J. T. Bowen, Philad.a 1845.

CANIS LUPUS, LINN. (VAR ATER.)
BLACK AMERICAN WOLF.
MALE.
⅓ Natural Size.

No. 14.

PLATE LXVIII.

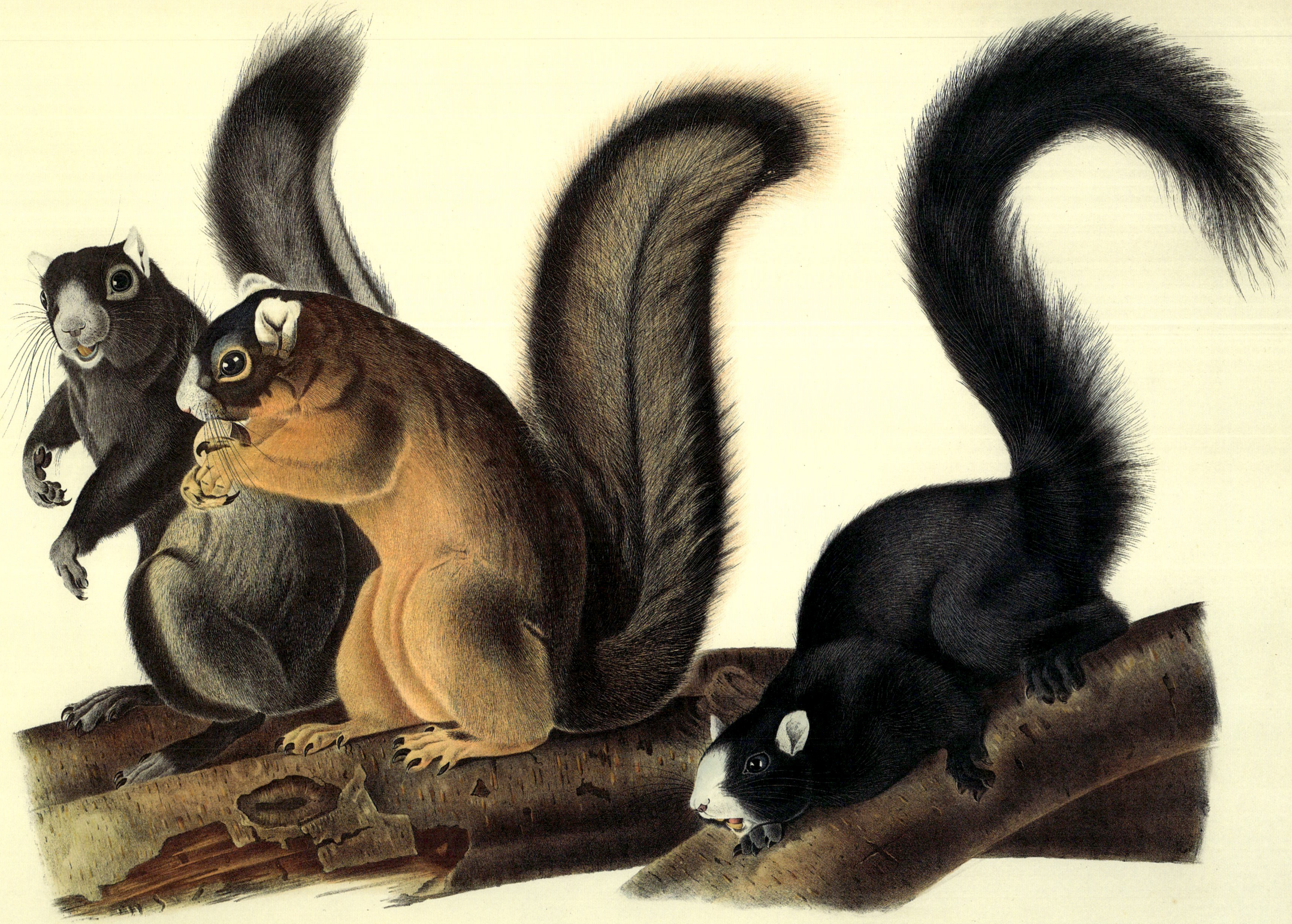

Drawn from Nature by J. J. Audubon F.R.S.F.L.S.

Lith. Printed & Col. by J. T. Bowen, Philad. 1845.

SCIURUS CAPISTRATUS, BOSC.

FOX SQUIRREL.

Natural Size.

PLATE LXIX

N° 14.

Drawn from Nature by J.J. Audubon. F.R.S.F.L.S.

CONDYLURA CRISTATA, LINN.

COMMON STAR-NOSE MOLE.

Natural Size.

Lith⁴ Printed & Col⁴ by J.T. Bowen. Philad⁴ 1845.

PLATE LXX.

Drawn from Nature by J.J.Audubon, F.R.S.F.L.S

Lith⁴ Printed & Col⁴ by J.T.Bowen, Phila⁴ 1845

SOREX PARVUS. SAY,

SAYS LEAST SHREW.

Natural Size.

PLATE LXXI.

CANIS LATRANS, SAY.
PRAIRIE WOLF.
MALES.
⅓ Natural Size.

Lith. Printed & Col. by J.T.Bowen, Philad.ª 1845.

Drawn from Nature by J.J.Audubon, F.R.S.F.L.S.

PLATE LXXII.

Nº 15

CANIS LUPUS, LINN (VAR ALBUS.)

WHITE AMERICAN WOLF.

MALE.

f. Natural Size

Drawn from Nature by J.J.Audubon F.R.S.F.L.S.

Lith.d Printed & Col.d by T. Bowen Philad.d 1844

No. 15.

PLATE LXXIII.

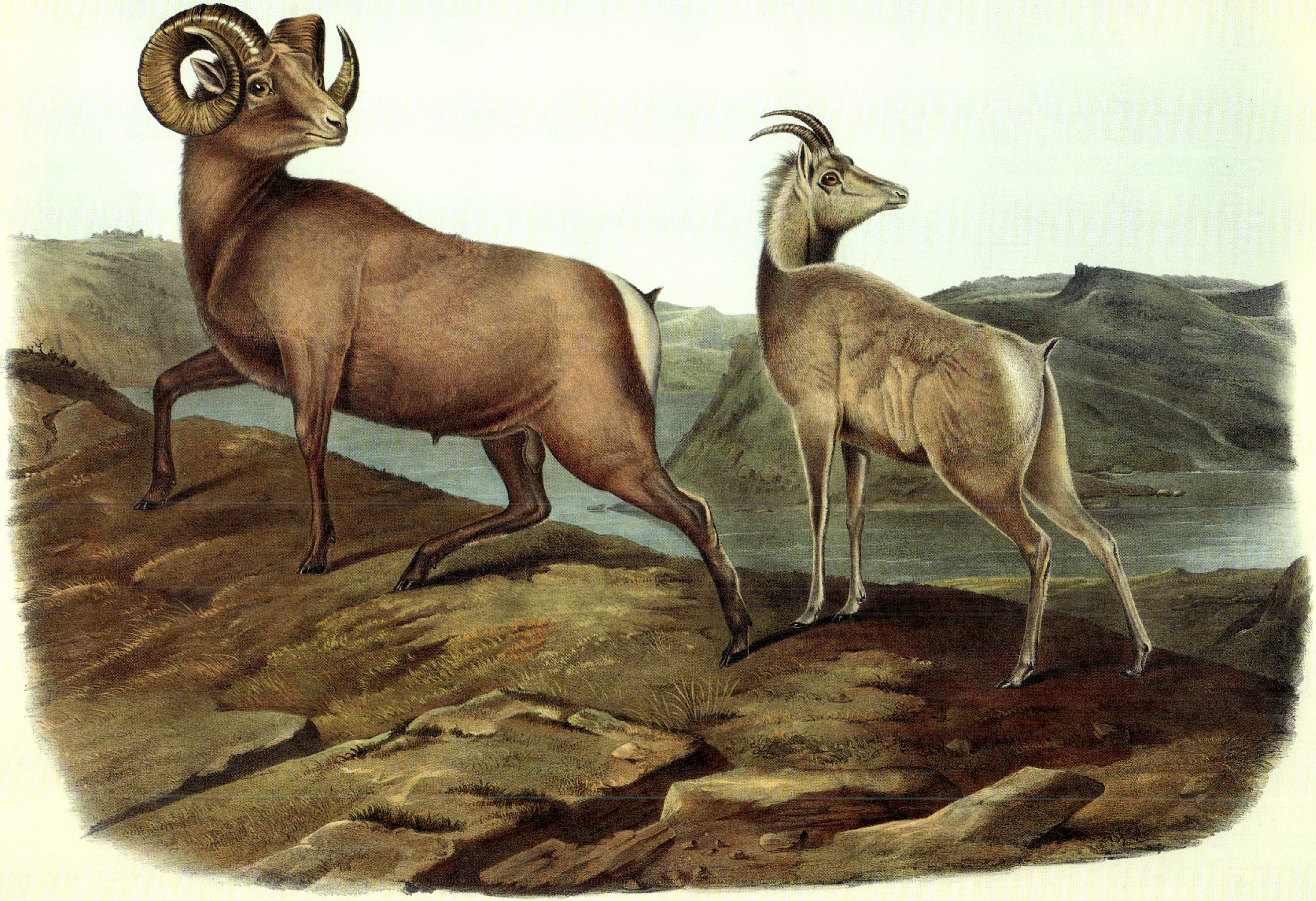

Drawn from Nature by J. J. Audubon, F.R.S. F.L.S.

Lith.d Printed & Col.d by J. T. Bowen, Philad.a 1845

OVIS MONTANA, DESM.

ROCKY MOUNTAIN SHEEP.

MALE & FEMALE.

¼ Natural Size.

PLATE LXXIV.

Drawn from Nature by J.J.Audubon, F.R.S.F.L.S.

Lith.d Printed & Col.d by J.T. Bowen, Philad.a 1845

SCALLOPS BREWERII, BACH.

BREWER'S SHREW MOLE.

MALE & FEMALE.

Natural Size.

PLATE LXXV.

No15.

Drawn from Nature by J. Audubon. F.R.S.F.L.S.

Lith.d Printed & Col.d by J. T. Bowen, Phila.a 1845

SOREX CAROLINENSIS. BACH.

CAROLINA SHREW.

MALE & FEMALE.

Natural Size.

PLATE LXXVI.

Drawn from Nature by J.J.Audubon, F.R.S.F.L.S.

Lith. Printed & Col.d by J.T. Bowen, Philad.a 1845.

CERVUS ALCES, LINN

MOOSE DEER.

OLD MALE. A YOUNG.

PLATE LXXVII.

Drawn from Nature by J.J.Audubon. F.R.S.F.L.S.

Lith Printed & Cold by J.T.Bowen, Philad 1845.

ANTILOPE AMERICANA. ORD.
PRONG-HORNED ANTELOPE.
MALES & FEMALE.

PLATE LXXVIII.

N° 16.

Lith Printed & Col. by J. T. Bowen. Phila 1845.

CERVUS MACROTIS, SAY.

BLACK - TAILED DEER.

FEMALE . SUMMER PELAGE.

PLATE LXXIX.

Drawn from Nature by J.J.Audubon F.R.S.F.L.S.

Lith⁴ Printed & Col⁴ by J.T.Bowen Philad.ᵃ 1845.

SPERMOPHILUS ANNULATUS, AUD. & BACH.

ANNULATED MARMOT SQUIRREL.

Natural Size.

PLATE LXXX.

N° 16

ARVICOLA PINETORUM, LECONTE
LECONTIS PINE MOUSE.
MALE & FEMALE.
Natural Size.

Drawn from Nature by J.J. Audubon, F.R.S.F.L.S.

Lith. Printed & Col. by J.T. Bowen, Philad.ª 1845.

PLATE LXXXI.

Drawn from Nature by J.J.Audubon, F.R.S.F.S.

Lith. Printed & Col.ª by J.T. Bowen, Philad.ª 1845.

CERVUS VIRGINIANUS, PENNANT.
COMMON AMERICAN DEER.
FAWN.
Natural Size.

PLATE LXXXII.

Drawn from Nature by J.J.Audubon, FRS.FLS.

Lith. Printed & Col. by J.T. Bowen, Phila. 1845.

CANIS LUPUS, LINN. VAR RUFUS.

RED TEXAN WOLF.

MALE.

PLATE LXXXIII.

Drawn from Nature by J.J. Audubon, F.R.S.F.L.S.

Lith? Printed & Col? by J. T. Bowen, Philad? 1843.

LAGOMYS PRINCEPS. RICHARDSON.

LITTLE CHIEF HARE.

Natural Size.

PLATE LXXXIV

Drawn from Nature by J. Audubon, FRS FLS

SPERMOPHILUS FRANKLINII, SABINE.

FRANKLINS MARMOT SQUIRREL.
MALE & FEMALE.
Natural Size.

Lith⁴ Printed & Col⁴ by J. T. Bowen, Phila⁴ 1846

PLATE LXXXV.

No 17.

MERIONES AMERICANUS, Barton.
JUMPING MOUSE.
MALE & FEMALE.
Natural Size.

Drawn from Nature by J.J.Audubon, F.R.S.F.L.S.

Lith Printed & col by J.T.Bowen Philad 1846.

PLATE LXXXVI.

Drawn from Nature by J.W. Audubon.

Lith. Printed & Col.d by T. Bowen, Philad.a 1846.

FELIS PARDALIS, LINN.
OCELOT, OR LEOPARD-CAT.
MALE.

PLATE LXXXVII.

VULPES FULVUS. (DESM.)

AMERICAN RED-FOX.

MALE.

PLATE. LXXXVIII.

Drawn from Nature by J.J.Audubon, FRS.F.L.S.

LEPUS ARTEMESIA, BACH.
WORM WOOD HARE.
MALE & FEMALE.
Natural Size.

Lithᵈ Printed & Colᵈ by J. T. Bowen, Philadᵃ 1846.

PLATE LXXXIX.

Nº 18.

SCIURUS SAYI. AUD. & BACH.

SAY'S SQUIRREL.

Natural Size.

Drawn from Nature by J.J.Audubon.F.R.S.F.L.S.

Lith.Printed & Col.d by J.T. Bowen, Philad.a 1846.

PLATE XC.

Drawn from Nature by J. Audubon, F.R.S.F.L.S.

MUS MUSCULUS, LINN.

COMMON MOUSE.

MALE, FEMALE & YOUNG.
Natural Size.

Lith.d Printed & Col.d by J. T. Bowen, Phila.a 1846.

PLATE XCI.

URSUS MARITIMUS, LINN.

POLAR BEAR.

MALE.

Drawn from Nature by J. W. Audubon.

Lith. Printed & Col.d by T. Bowen, Phila.d 1846.

PLATE XCII.

Drawn from Nature by J.J. Audubon, F.R.S. F.L.S.

Lithd Printed & Cold by J. T. Bowen, Philada 1846.

LYNX RUFUS., VAR MACULATUS, HORSFIELD & VIGORS.

TEXAN LYNX.

FEMALE.

No. 19.

PLATE XCIII.

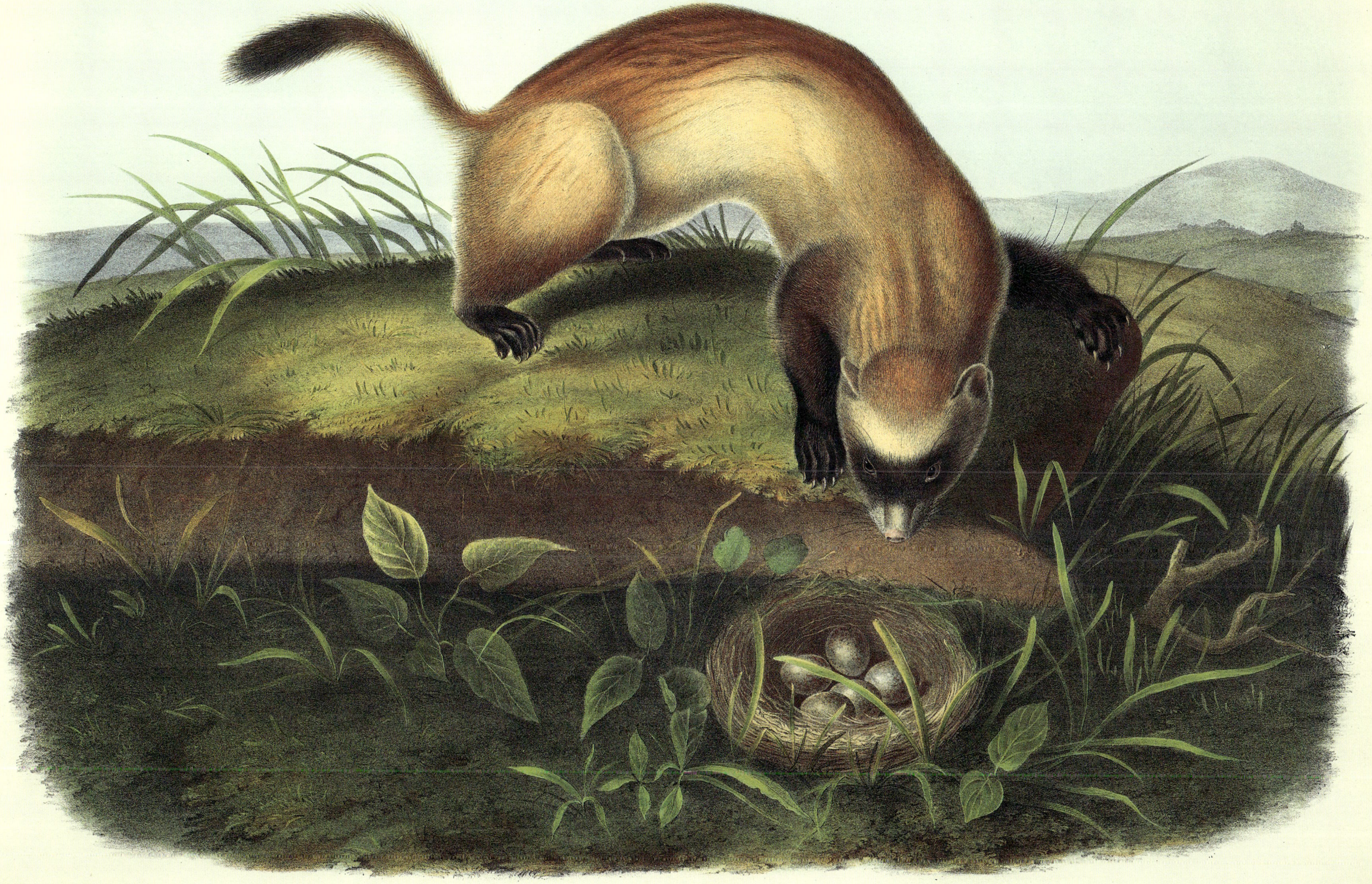

Drawn from Nature by J. W. Audubon

Lith.d Printed & Col.d by J. T. Bowen, Philad.a 1846

PUTORIUS NIGRIPES, AUD. & BACH.

BLACK FOOTED FERRET.

Natural Size.

PLATE XCIV.

LEPUS NUTTALLII, BACH.

NUTTALL'S HARE.

MALES.

Natural Size.

Drawn from Nature by J. W. Audubon.

Lith. Printed & Col. by T. Bowen, Philad. 1846.

PLATE XCV.

No. 19.

MUS AUREOLUS, AUD. & BACH.
ORANGE COLORED MOUSE.
MALE & FEMALE.
Natural Size.

Drawn from Nature by J. W. Audubon.

Lith. Printed & Col.d by J. T. Bowen, Philad.a 1846.

No 20.

PLATE XCVI.

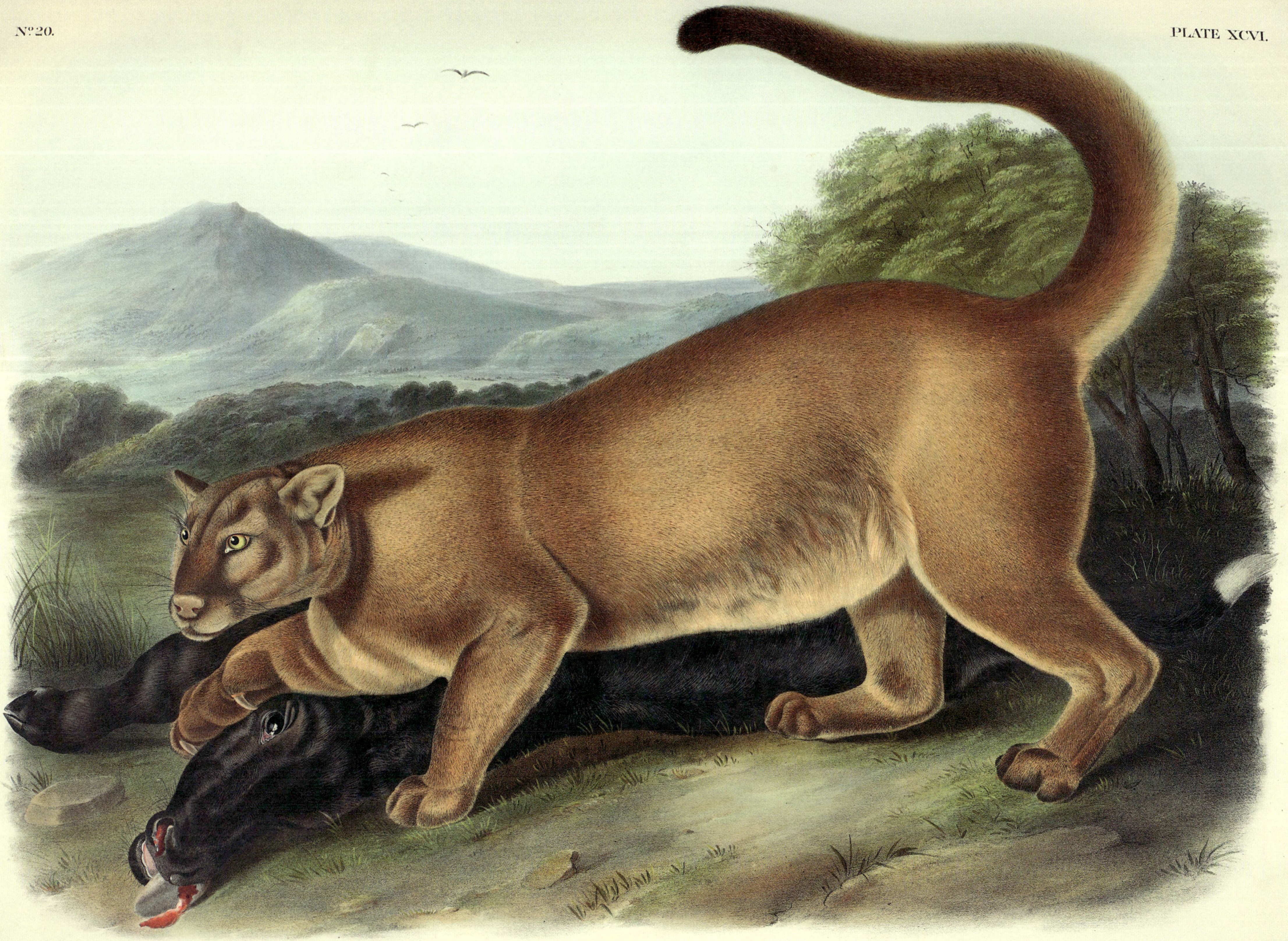

Drawn from Nature by J.W. Audubon.

FELIS CONCOLOR, LINN.

THE COUGAR.

MALE.

Lith^d Printed & Col^d by J.T. Bowen, Philad^a 1846.

PLATE XCVII.

Lith⁴ Printed & Col⁴ by J. T. Bowen, Philad⁴ 1846.

FELIS CONCOLOR, LINN.

THE COUGAR.

FEMALE & YOUNG.

Drawn from Nature by J. W. Audubon.

Drawn from Nature by J.J.Audubon, F.R.S.F.L.S.

Lith.d Printed & Col.d by J. T. Bowen, Philad.a 1846.

BASSARIS ASTUTA, LICHT.
RING-TAILED BASSARIS.
Natural Size.
MALE.

PLATE XCIX.

Drawn from Nature by J.J. Audubon FRSFLS

Lith. Printed & Col.d by T. Bowen, Philad.t 1846.

SPERMOPHILUS LUDOVICIANUS, ORD.

PRAIRIE DOG – PRAIRIE MARMOT SQUIRREL.

Natural Size.

1 MALE. 2 FEMALE. 3 YOUNG.

PLATE C.

Drawn from Nature by J.W. Audubon.

Lith.d Printed & Col.d by J.T. Bowen, Phila.d 1846.

MUS MISSOURIENSIS, AUD. & BACH.

MISSOURI MOUSE.

Natural Size.

No. 21. PLATE CI.

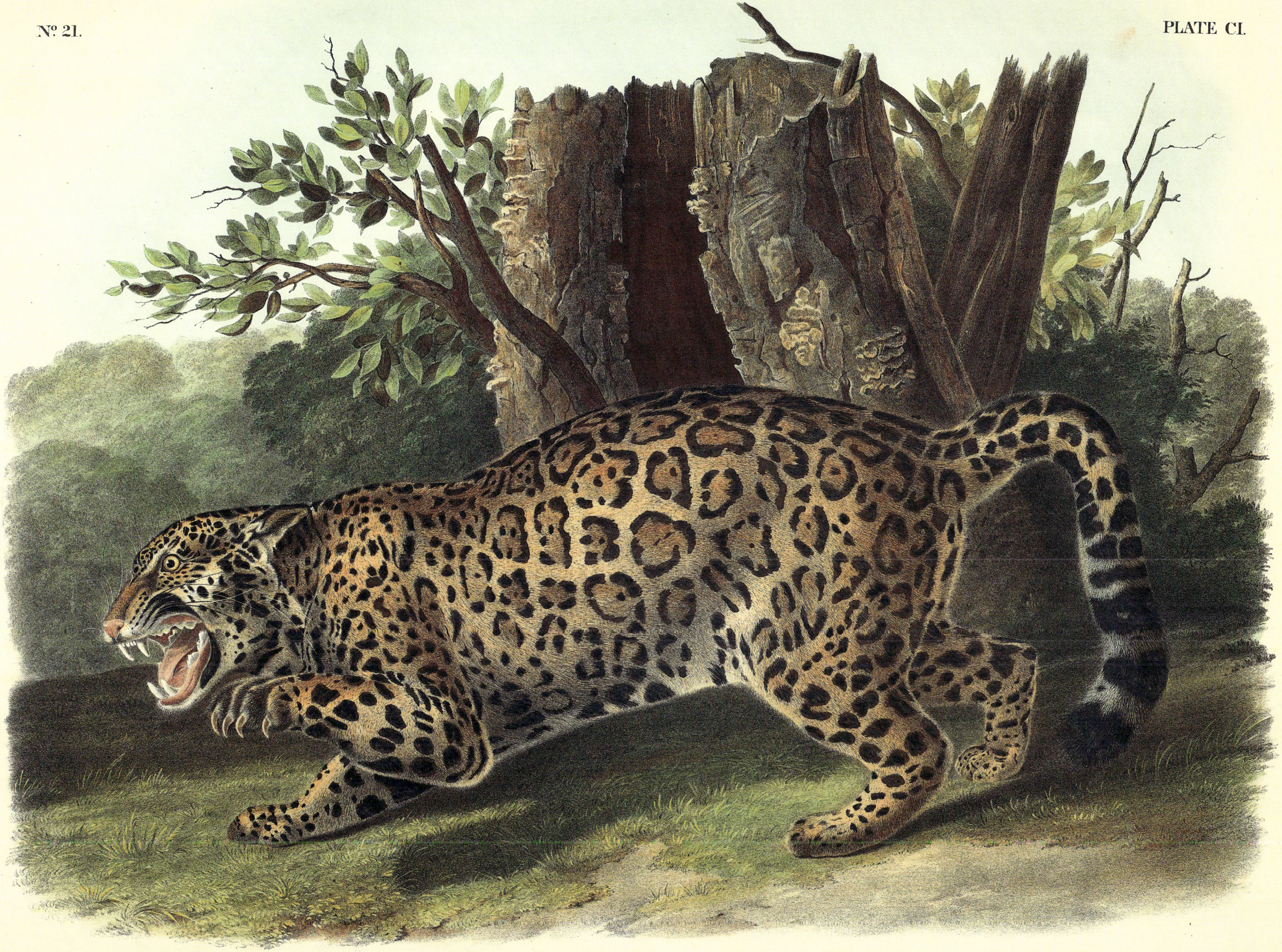

Drawn from Nature by J.J. Audubon F.R.S.F.L.S.

FELIS ONCA, LINN.

THE JAGUAR.

FEMALE

Lith^d. Printed & Col^d. by J.T. Bowen, Philad^a. 1846.

Drawn from Nature by J.W. Audubon.

Lith⁴ Printed & Col⁴ by J.T. Bowen, Philad⁴ 1846.

MEPHITIS MACROURA, Licht.

LARGE TAILED SKUNK.

MALE.

Natural Size.

No. 21.

PLATE CIII.

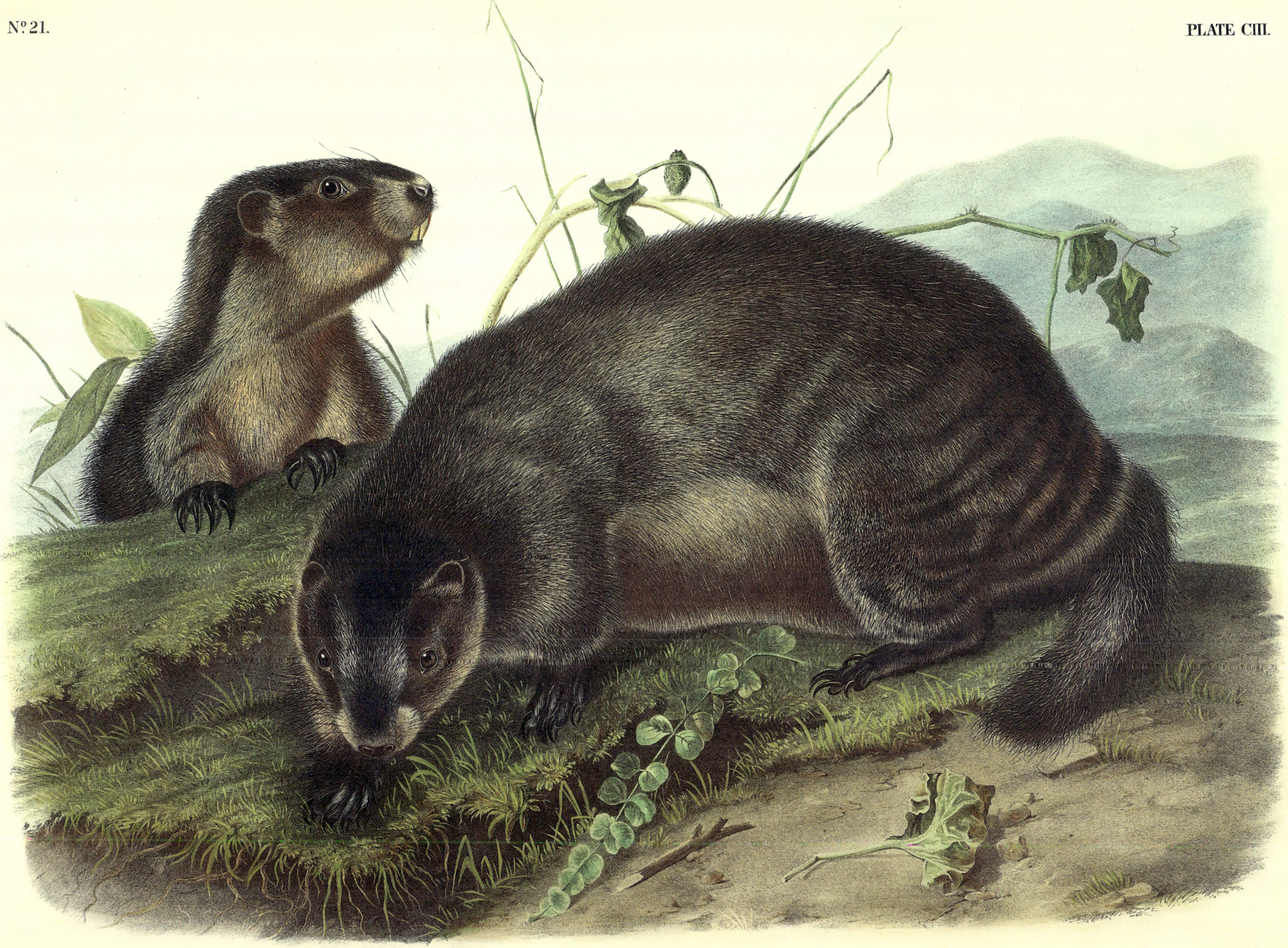

Drawn from Nature by J. W. Audubon.

ARCTOMYS PRUINOSUS, PENNANT.

HOARY MARMOT - THE WHISTLER.

Lith.d Printed & Col.d by J. T. Bowen, Philad.a 1846.

PLATE CIV.

Drawn from Nature by J.W. Audubon

Lith. Printed & Col.d by J.T. Bowen, Philad.a 1847.

SCIURUS COLLIÆI, RICH.

COLLIES SQUIRREL

Natural Size.

No. 21. PLATE CV.

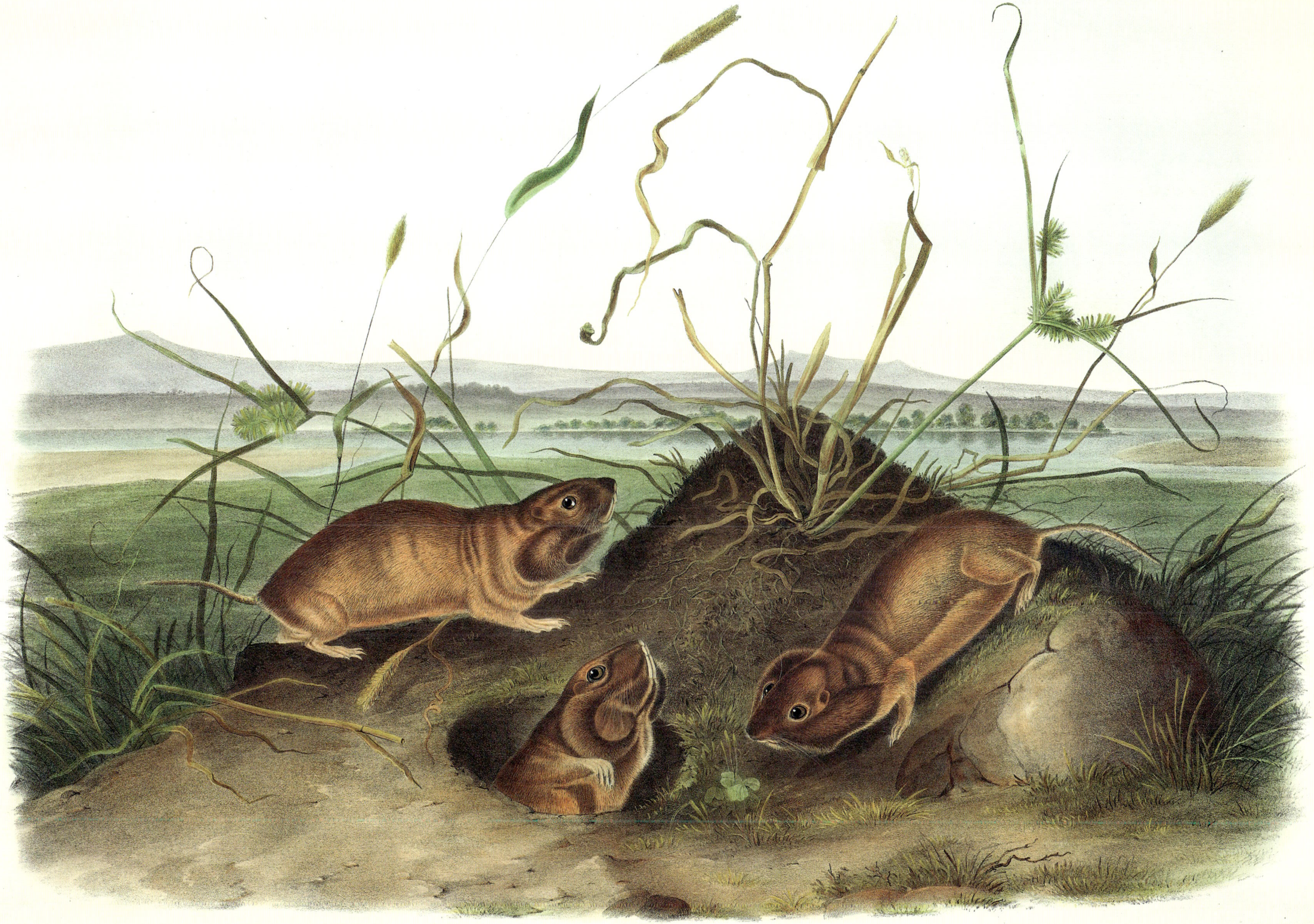

Drawn from Nature by J.W. Audubon.

GEOMYS DOUGLASSII, RICHARDSON

COLUMBIA POUCHED RAT.

Natural Size.

Lith.d Printed & Col.d by J.T. Bowen, Philad.a 1846.

PLATE CVI.

No. 22.

Drawn from Nature by J. J. Audubon, F.R.S.F.L.S.

CERVUS RICHARDSONII, AUD. & BACH.
COLUMBIAN BLACK-TAILED DEER.
Males.

Lith Printed & Cold by J. T. Bowen, Philad.ᵃ 1847.

Nº 22.

PLATE CVII.

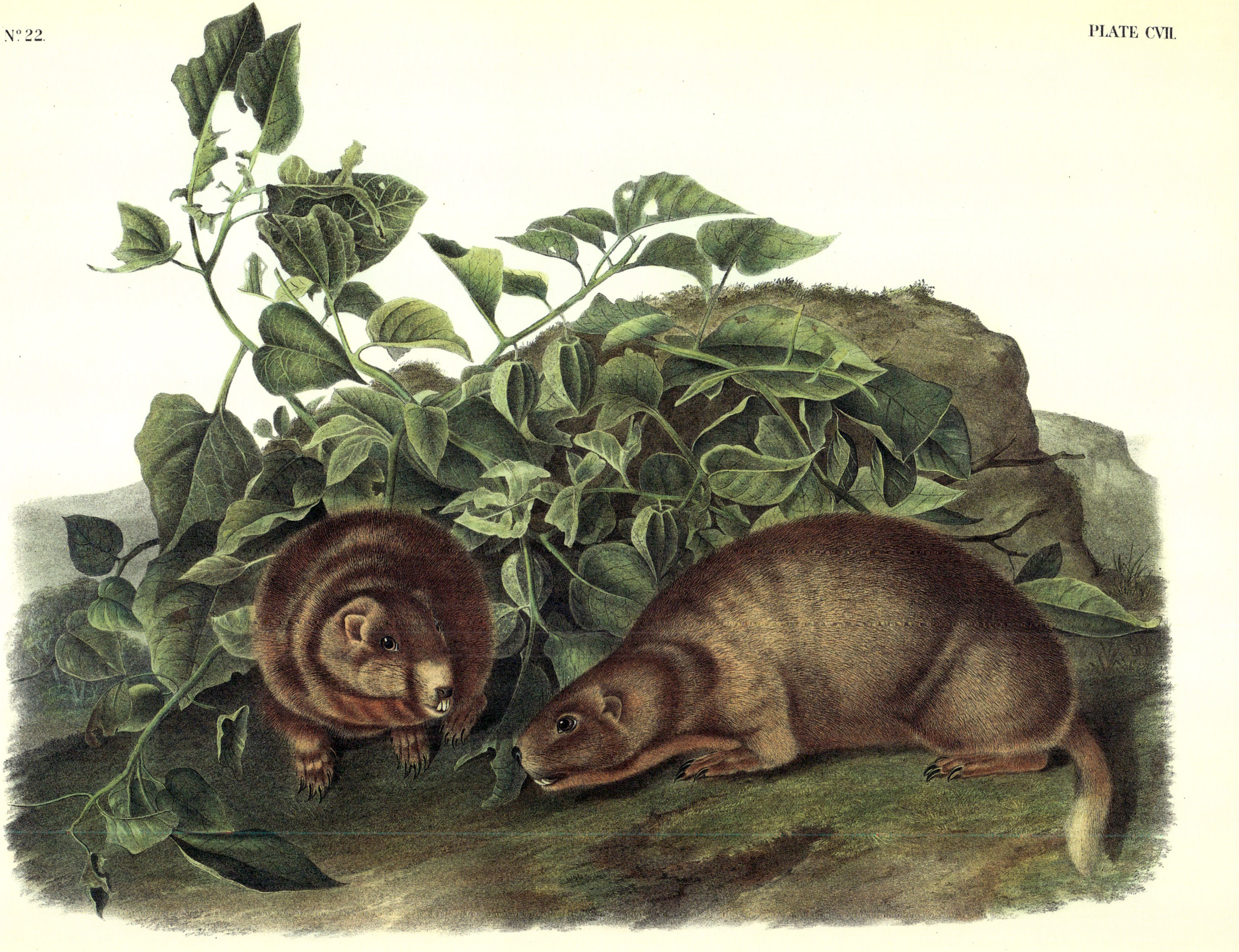

Drawn from Nature by J. W. Audubon.

Lith.d, Printed & Col.d by J. T. Bowen, Philad.a 1847.

ARCTOMYS LEWISII, AUD. & BACH.

LEWIS' MARMOT.

Natural Size.

PLATE CVIII.

LEPUS BACHMANI. WATERHOUSE.

BACHMAN'S HARE.

Natural Size.

Drawn from Nature by J. W. Audubon.

Lith. Printed & Col.ᵈ by J. T. Bowen, Philad.ᵃ 1847.

PLATE CIX.

Nº 22.

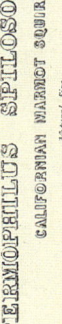

Lith Printed & Col. by J. T. Bowen. Philad.ª 1847

Drawn from Nature by J.J.Audubon. F.R.S.F.L.S.

SPERMOPHILUS SPILOSOMA, BENNET.
CALIFORNIAN HARVEST SQUIRREL.
Natural Size.

PLATE CX.

N°.22.

Drawn from Nature by J. W. Audubon.

Lith. Printed & Col.d by J. T. Bowen, Philad.a 1847.

PSEUDOSTOMA TALPOIDES, RICH.

MOLE SHAPED POUCHED RAT.

Natural Size.

PLATE CXL.

Drawn from Nature by J.W.Audubon.

OVIBOS MOSCHATUS, GMEL.
MUSK OX

Lith.ª Printed & Col.ª by J.T.Bowen, Philad.ª 1847.

PLATE CXII.

Drawn from Nature by J. W. Audubon.

Lith Printed & Col by J. T. Bowen, Philad.º 1847.

LEPUS CALIFORNICUS, GRAY.

CALIFORNIAN HARE.

Natural Size.

PLATE CXIII.

Drawn from Nature by J.W.Audubon.

CANIS FAMILIARIS, LINN. VAR. BOREALIS, DESM.
ESQUIMAUX DOG.

Lith.Printed & Col.d by J.T.Bowen, Philad.a 1847.

PLATE CXIV

Drawn from Nature by J. W. Audubon

Lith Printed & Cold by J. T. Bowen, Philad. 1847

SPERMOPHILUS LATERALIS, SAY.

SAY'S MARMOT SQUIRREL.

Natural Size.

No. 23.

PLATE CXV.

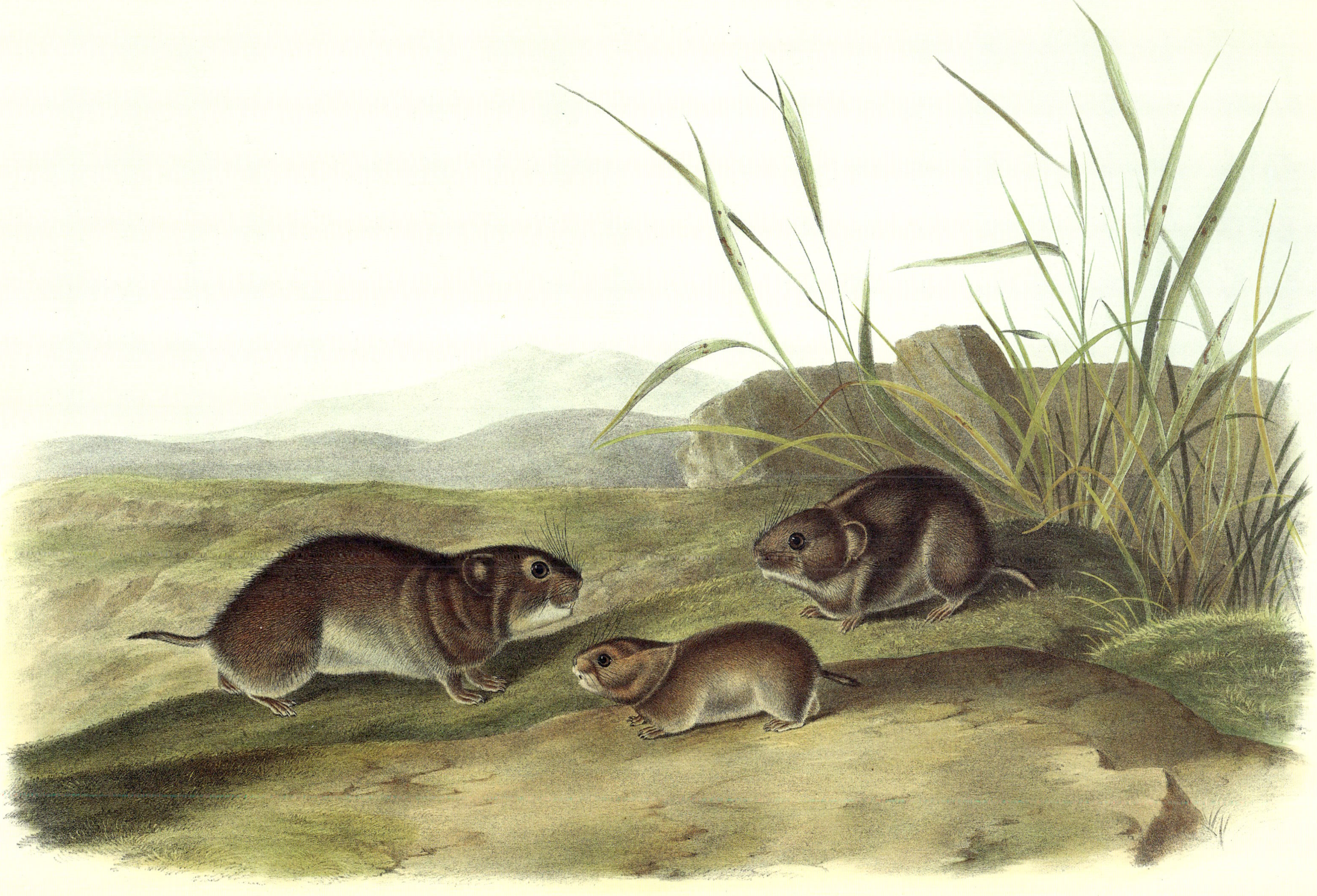

Drawn from Nature by J. W. Audubon

Lith.d Printed & Col.d by J. T. Bowen, Philad.a 1847

ARVICOLA XANTHOGNATHUS, LEACH.

YELLOW CHEEKED MEADOW MOUSE.

Natural Size.

PLATE CXVI.

Lith. Printed & Col. by J.T. Bowen, Philad. 1847.

VULPES FULVUS, DESM. VAR ARGENTATUS, RICH.

AMERICAN BLACK OR SILVER FOX.

Natural Size.

No. 24.

Drawn from Nature by J. W. Audubon.

PLATE CXVII.

Drawn from Nature by J.W.Audubon.

Lith. Printed & Col.d by J.T. Bowen, Philad.a

SCIURUS NIGRESCENS, BENNETT.

DUSKY SQUIRREL.

MALE.

Natural Size.

PLATE CXVIII.

CERVUS LEUCURUS, DOUGLASS

LONG-TAILED DEER.

Male.

Drawn from Nature by J.W. Audubon.

Lith.d Printed & Col.d by J. T. Bowen, Philad.a 1847

No. 24

PLATE CXIX.

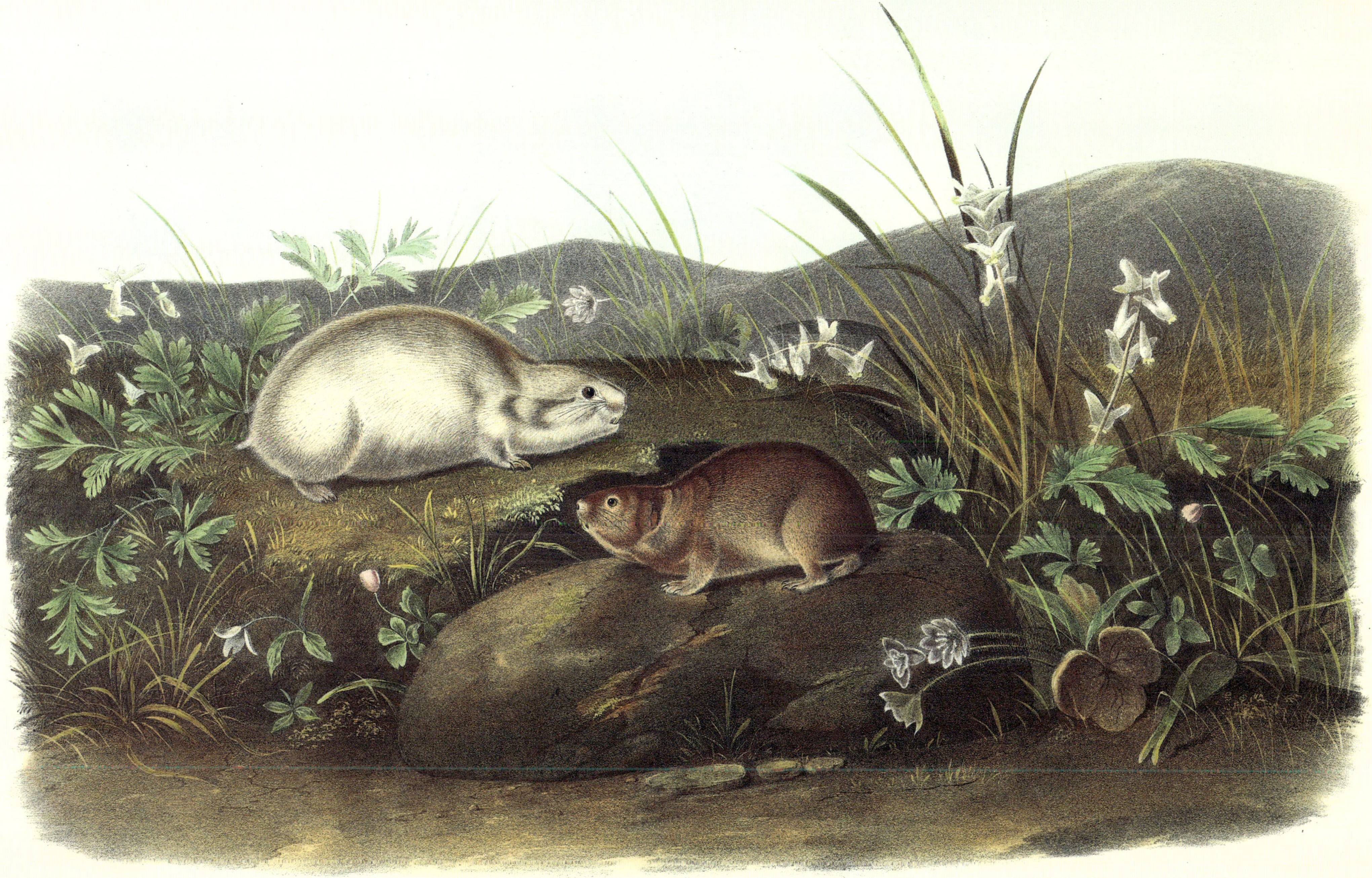

Drawn from Nature by J. W. Audubon.

MYODES HUDSONIUS, RICH.

HUDSON'S BAY LEMMING

Natural Size

Lith.d Printed & Col.d by J. T. Bowen, Philad.a 1847

PLATE CXX.

Nº 24.

Fig. 1. GEORYCHUS HELVOLUS, RICH. | Fig.2 & 3. GEORYCHUS TRIMUCRONATUS, RICH.

TAWNY LEMMING.

Natural Size.

BACK'S LEMMING.

Natural Size.

Drawn from Nature by J.J.Audubon. FRS.FLS.

Lith.Printed & Col.d by J.T.Bowen. Phila.d 1847.

PLATE CXXI.

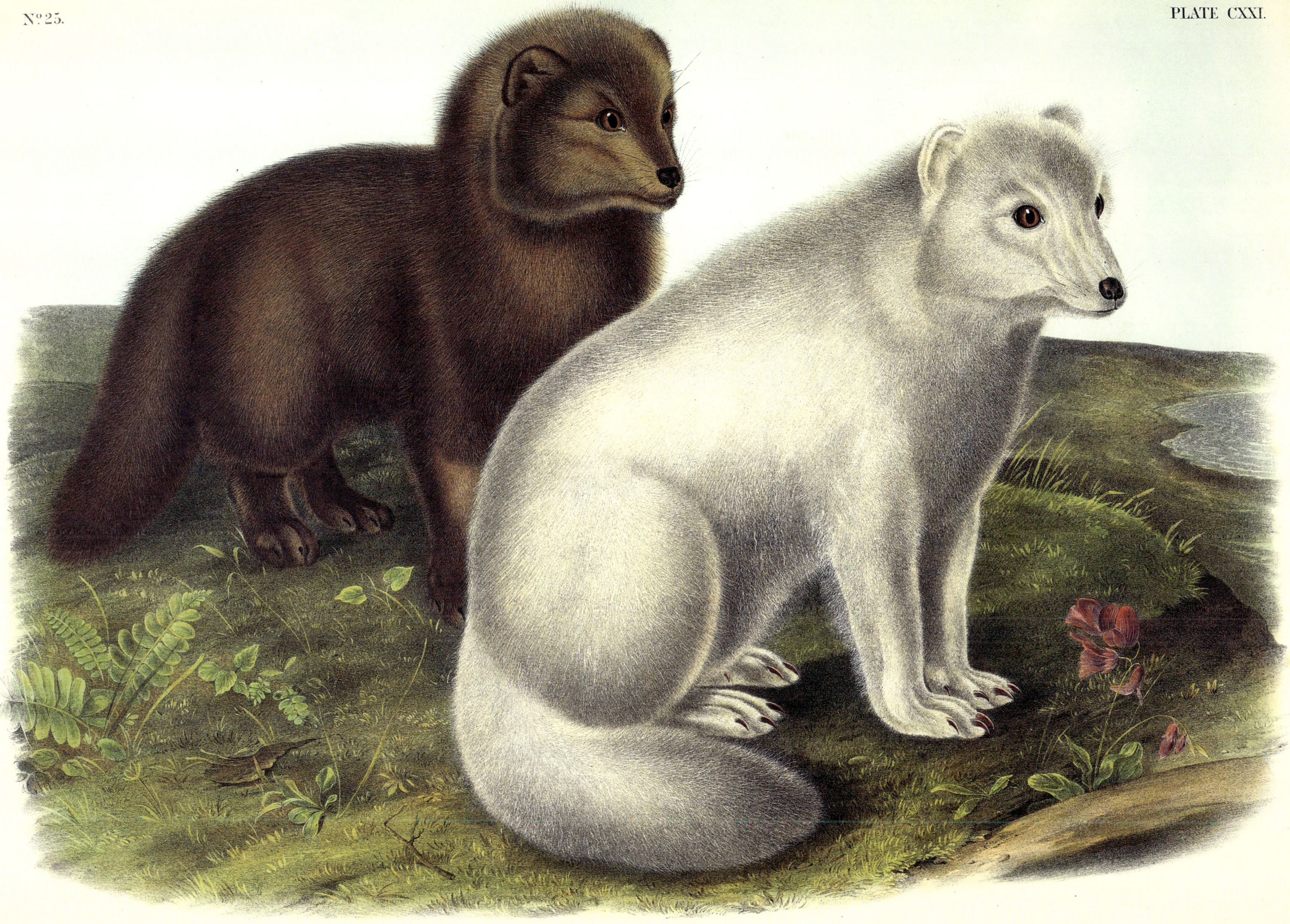

VULPES LAGOPUS, LINN.

ARCTIC FOX.

WINTER & SUMMER PELAGE.

No. 25.

PLATE CXXII.

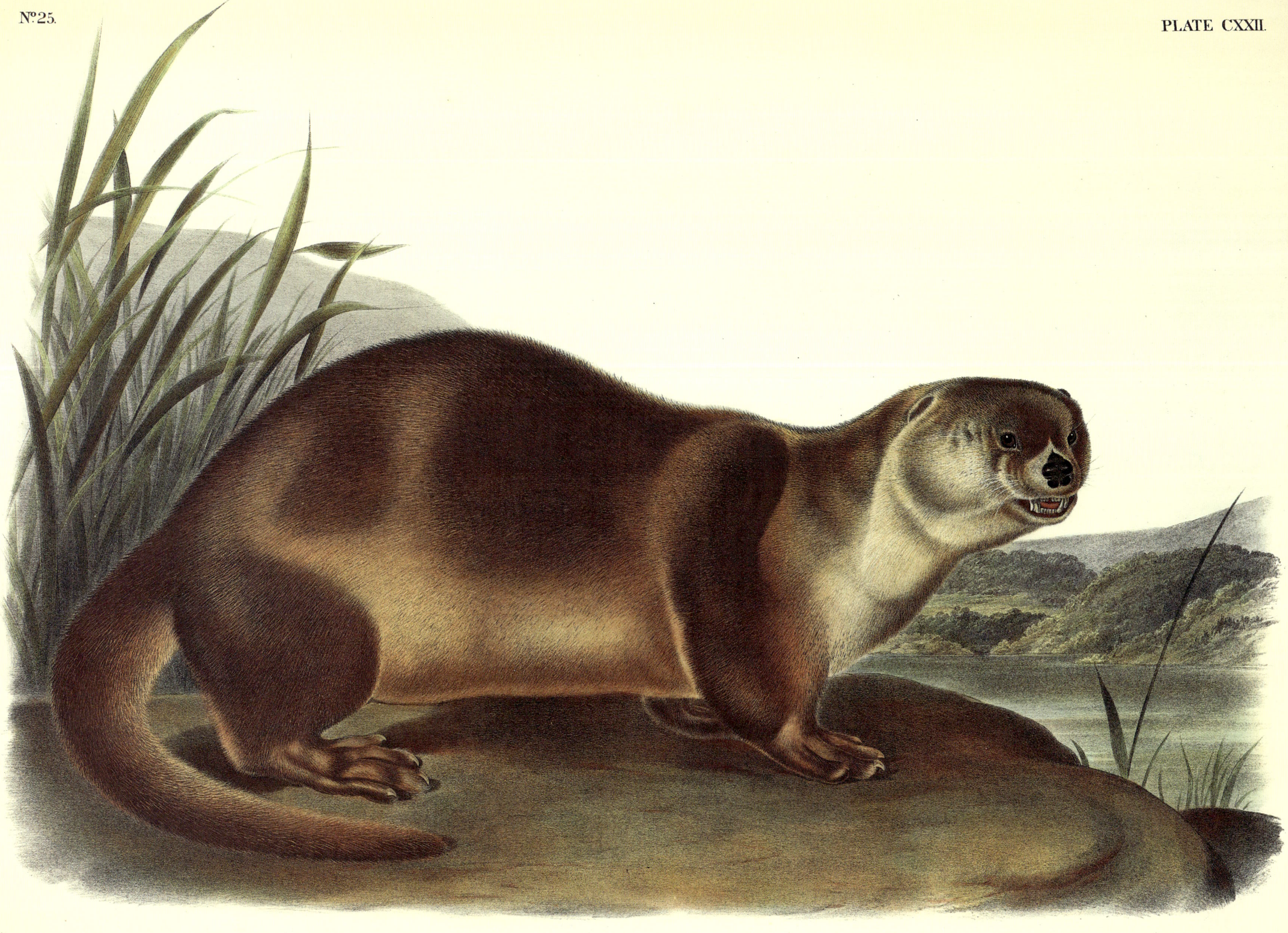

Drawn from Nature by J. W. Audubon.

LUTRA CANADENSIS, SABINE, VAR.

LATAXINA MOLLIS, GRAY.

CANADA OTTER–MALE.

Lith.d Printed & Col.d by J. T. Bowen, Philad.a 1847.

No. 25.

PLATE CXXIII.

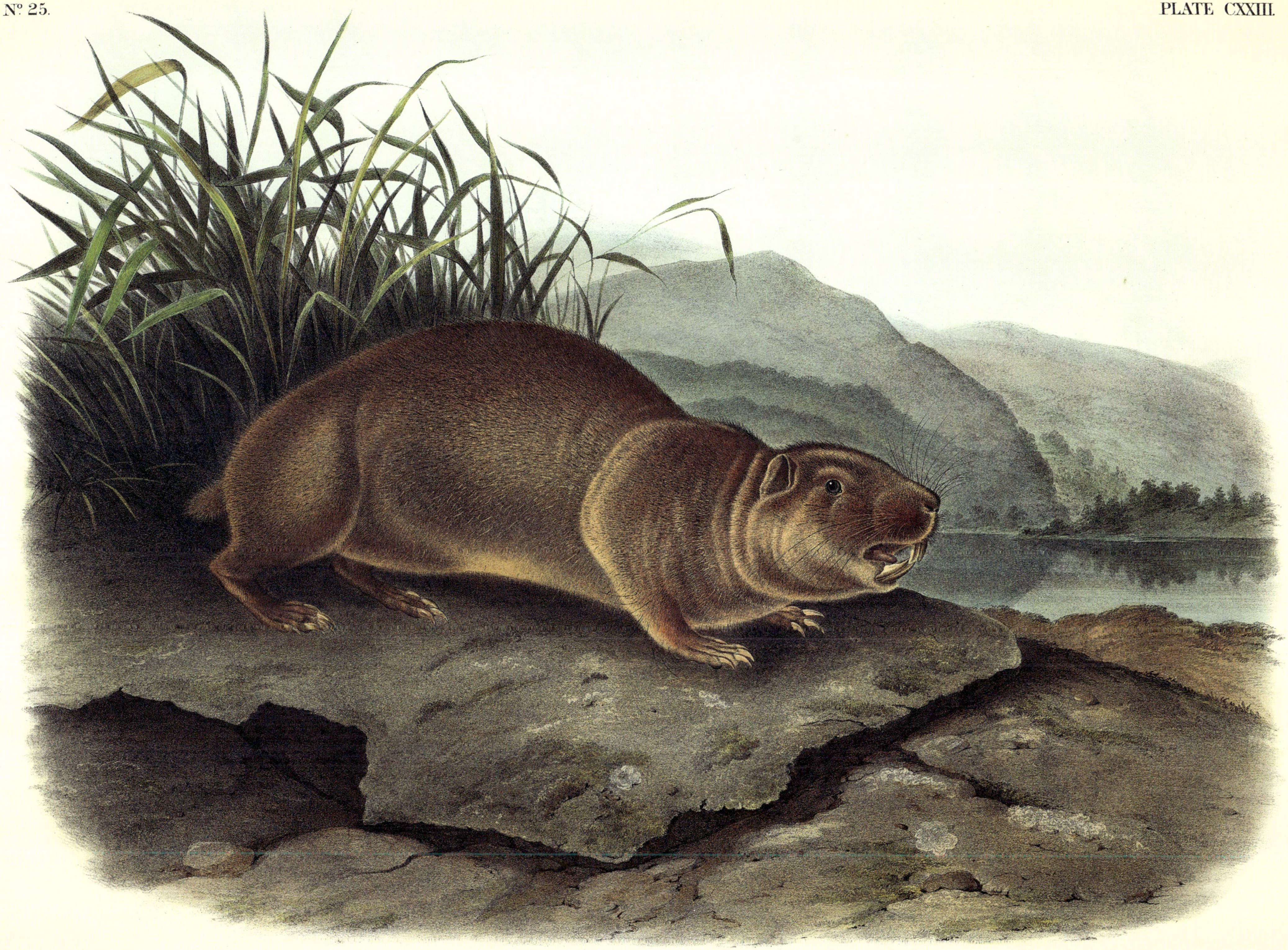

Drawn from Nature by J. W. Audubon

APLODONTIA LEPORINA, RICH.

THE SEWELLEL.

MALE.

Natural Size.

Lith. Printed & Col. by J. T. Bowen, Philada. 1847

PLATE CXXIV.

No 25.

Drawn from Nature by J. W. Audubon.

Lith. Printed & Col.d by J.T.Bowen, Philad.a 1847.

SPERMOPHILUS MEXICANUS LICHT.

MEXICAN MARMOT - SQUIRREL.

MALE.

Natural size.

No. 25. PLATE CXXV.

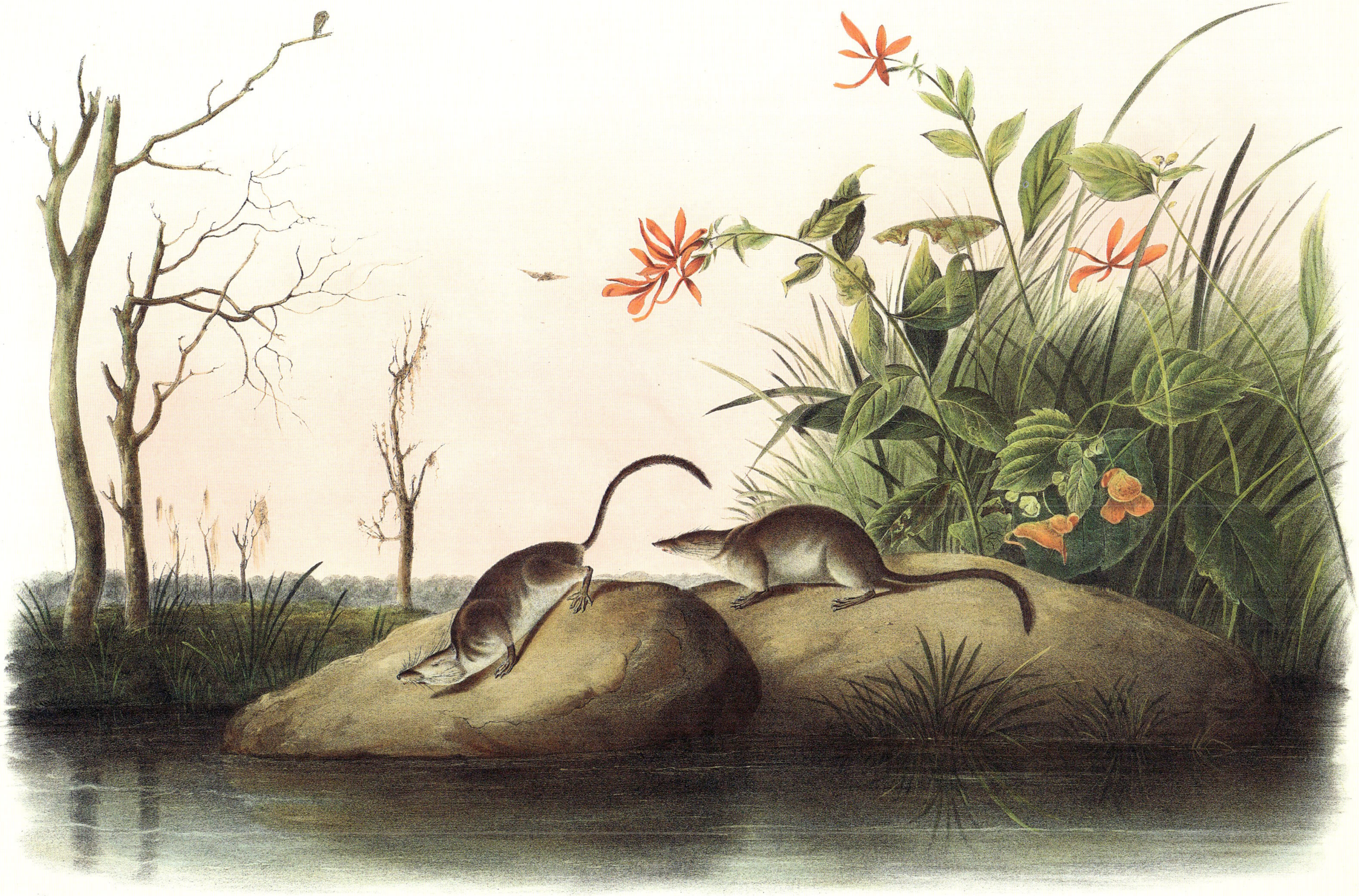

Drawn from Nature by J.W. Audubon

SOREX PALUSTRIS, RICH.

AMERICAN MARSH SHREW.

MALES.

Natural Size.

Lith.d Printed & Col.d by J.T. Bowen, Philad.a 1847.

PLATE CXXVI.

N°.26.

TARANDUS FURCIFER. AGASSIZ.

CARIBOU OR AMERICAN REIN-DEER.

MALES.

1 Summer pelage, 2 winter pelage.

Drawn from Nature by J. W. Audubon.

Lith d Printed & Col d by J. T. Bowen, Philad a 1847.

PLATE CXXVII.

URSUS AMERICANUS, PALLAS.

VAR. CINNAMOMUM, AUD & BACH.

CINNAMON BEAR.

MALE & FEMALE.

Drawn from Nature by J. W. Audubon.

Lith. Printed & Col.d by J. T. Bowen, Phila.d 1847.

PLATE CXXVIII.

Drawn from Nature, by J.W. Audubon.

Lith.d Printed & Col.d by J. T. Bowen, Philad.ª 1847.

CAPRA AMERICANA, BLAINVILLE.

ROCKY MOUNTAIN GOAT.

MALE & FEMALE.

PLATE CXXIV.

ARVICOLA BOREALIS. RICH.

NORTHERN MEADOW-MOUSE.

Natural Size.

Drawn from Nature by J.W.Audubon.

Lith. Printed & Col. by J.T Bowen, Philad.ª 1848.

PLATE CXX.

Nº 26.

DIPODOMYS PHILLIPSII, GRAY.

POUCHED JERBOA MOUSE.

MALES.

Natural Size.

Drawn from Nature by J. W. Audubon.

Lith. Printed & Col. by J. T. Bowen, Philad. 1847.

No. 27.

PLATE CXXXI.

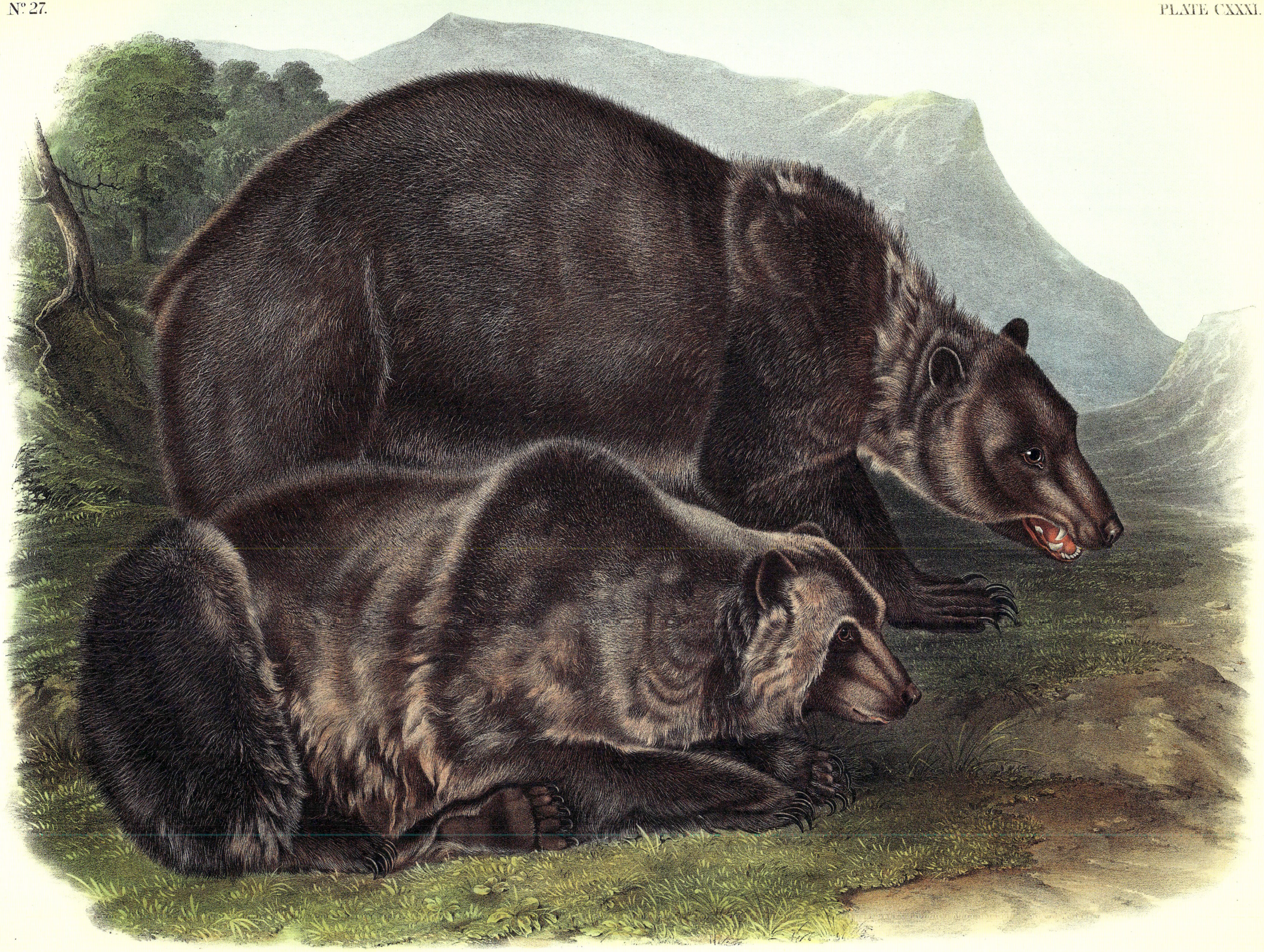

Drawn from Nature by J. W. Audubon.

URSUS FEROX, LEWIS & CLARK.

GRIZZLY BEAR.

MALES

Lith. Printed & Col. by J. T. Bowen, Philada. 1848.

No. 27.

PLATE CXXXII.

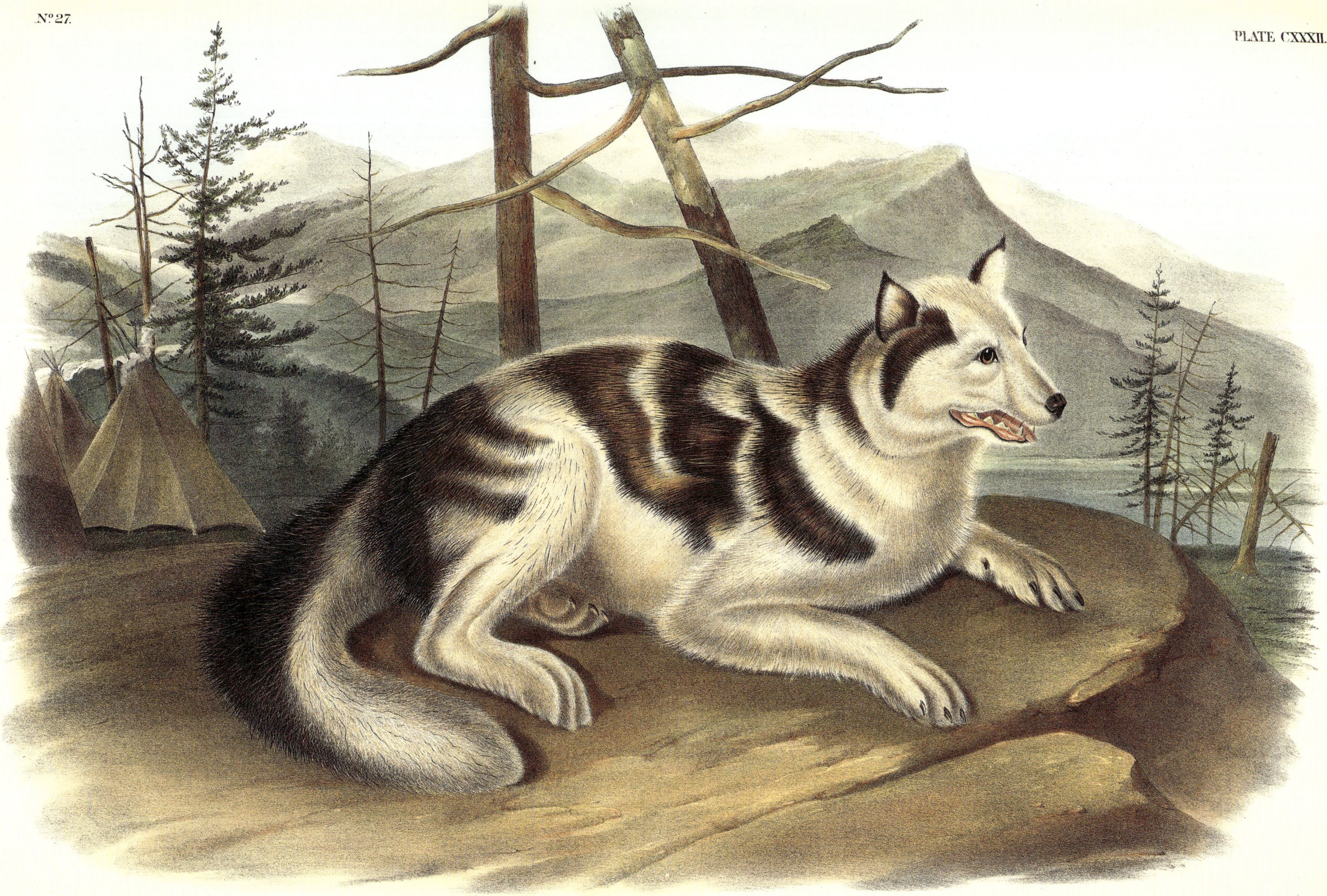

CANIS FAMILIARIS, LINN. (VAR LAGOPUS.)

HARE-INDIAN DOG.

MALE.

Lith.d Printed & Col.d by J. T. Bowen, Philad.a 1848.

PLATE CXXXIII.

LEPUS TEXIANUS, AUD. & BACH.

TEXIAN HARE.

MALE.

Natural Size.

Drawn from Nature by J. W. Audubon.

Lith.d Printed & Col.d by J.T.Bowen, Phila.d 1848

PLATE CXXXIV.

ARCTOMYS FLAVIVENTER, BACH.

YELLOW BELLIED MARMOT.

MALE.

Natural Size.

Drawn from Nature by J.W.Audubon.

Lith.d Printed & Col.d by J.T Bowen, Philad.ª 1843.

PLATE CXXXV

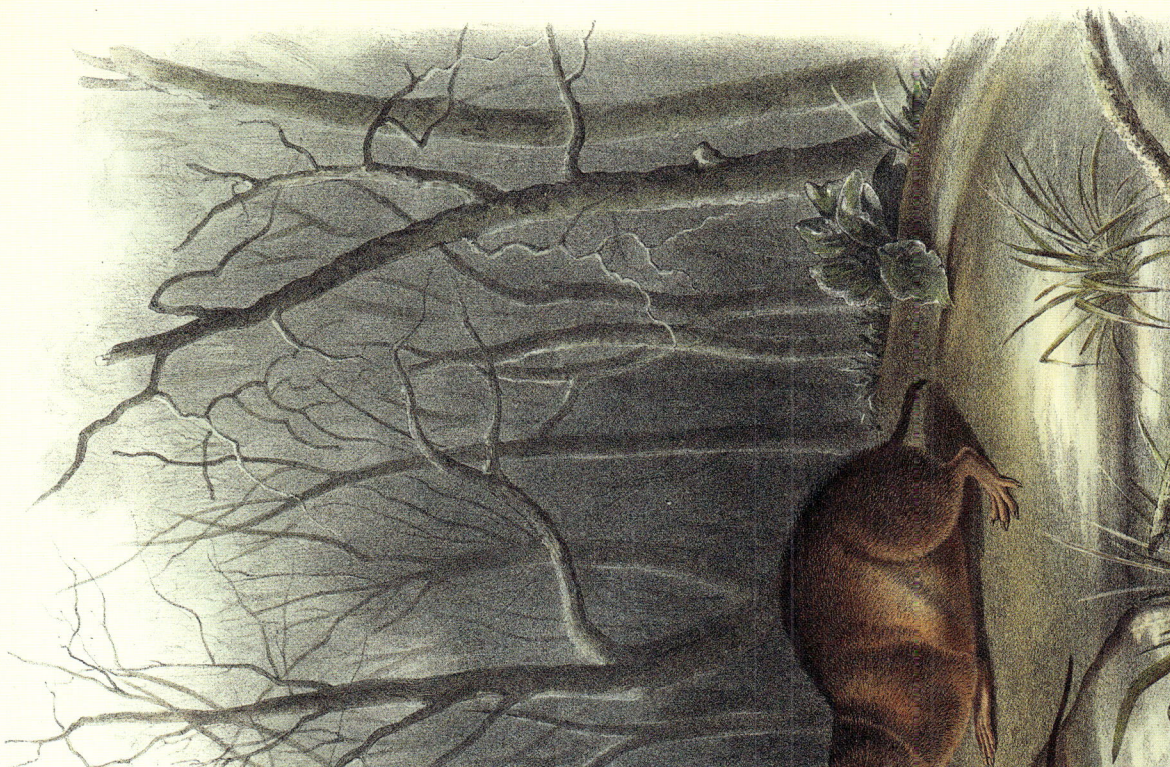

Drawn from Nature by J.W. Audubon.

Lith. Printed & Col₫ by J.T. Bowen, Philad™ 1843.

ARVICOLA RICHARDSONII, AUD. & BACH.

RICHARDSON'S MEADOW MOUSE.

Natural Size.

PLATE CXXXVI.

N.° 28.

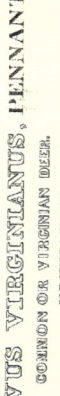

CERVUS VIRGINIANUS, PENNANT.

COMMON OR VIRGINIAN DEER.

OLD MALE & FEMALE.

Drawn from Nature by J.W. Audubon.

Lith.d Printed & Col.d by J.T. Bowen, Phila.d 1848.

PLATE CXXXVII.

ENHYDRA MARINA, ERXLEBEN.
SEA OTTER.
YOUNG MALE.

Drawn from Nature by J W Audubon.

Lith: Printed & Col: by J.T.Bowen, Philad: 1848.

PLATE CXXXVIII.

MUSTELA MARTES. LINN.
PINE MARTEN.
MALE & FEMALE. WINTER PELAGE.
Natural Size.

Lith. Printed & Col. by J.T. Bowen, Phil.ᵈ 1848.

No. 28.

PLATE CXXXIX.

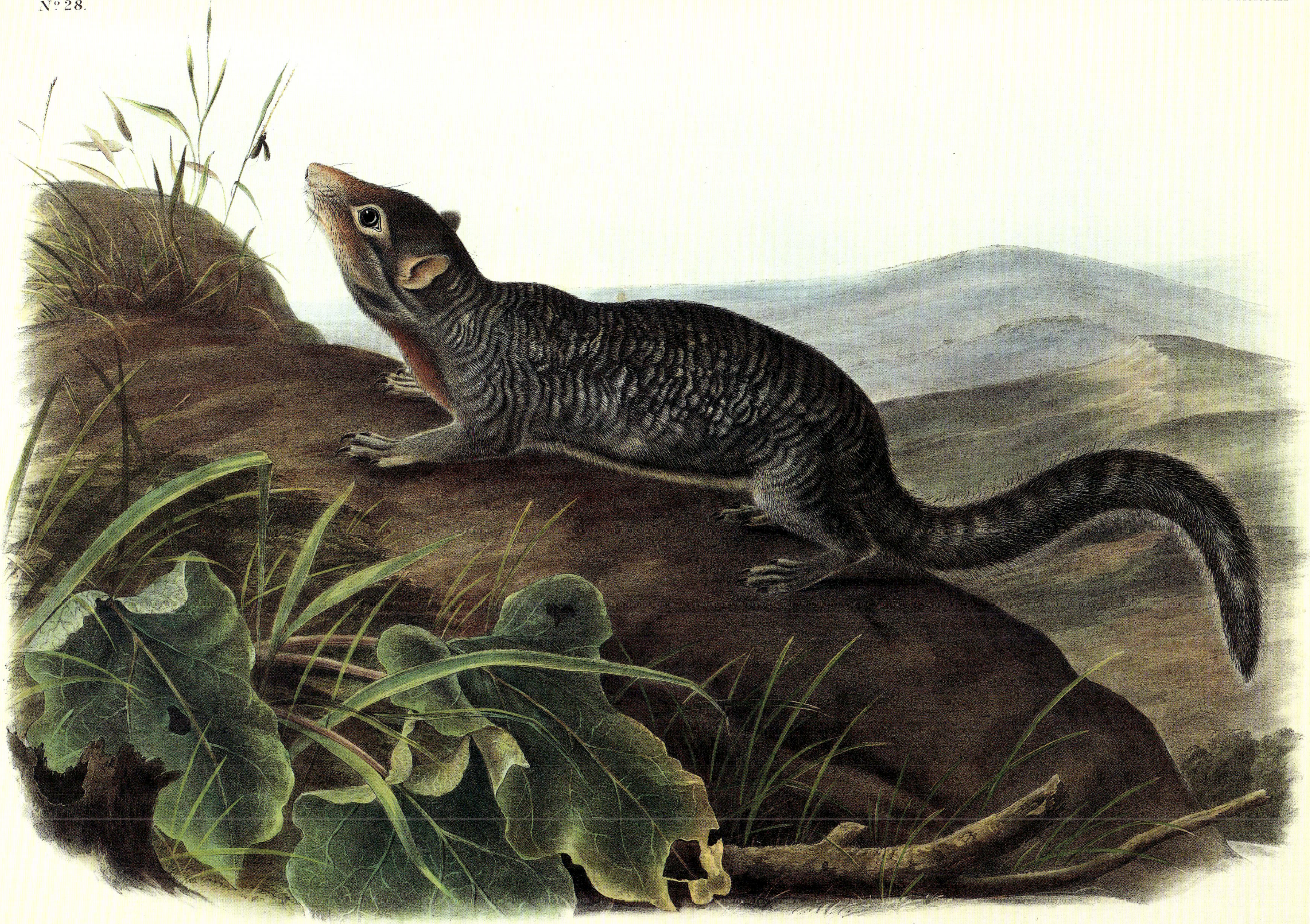

Drawn from Nature by J. W. Audubon

Lith. Printed & Col.d by J. T. Bowen, Philad.a 1848

SPERMOPHILUS MACROURUS, BENNETT.

LARGE-TAILED SPERMOPHILE.

MALE.

Natural Size

N° 28

PLATE CXL.

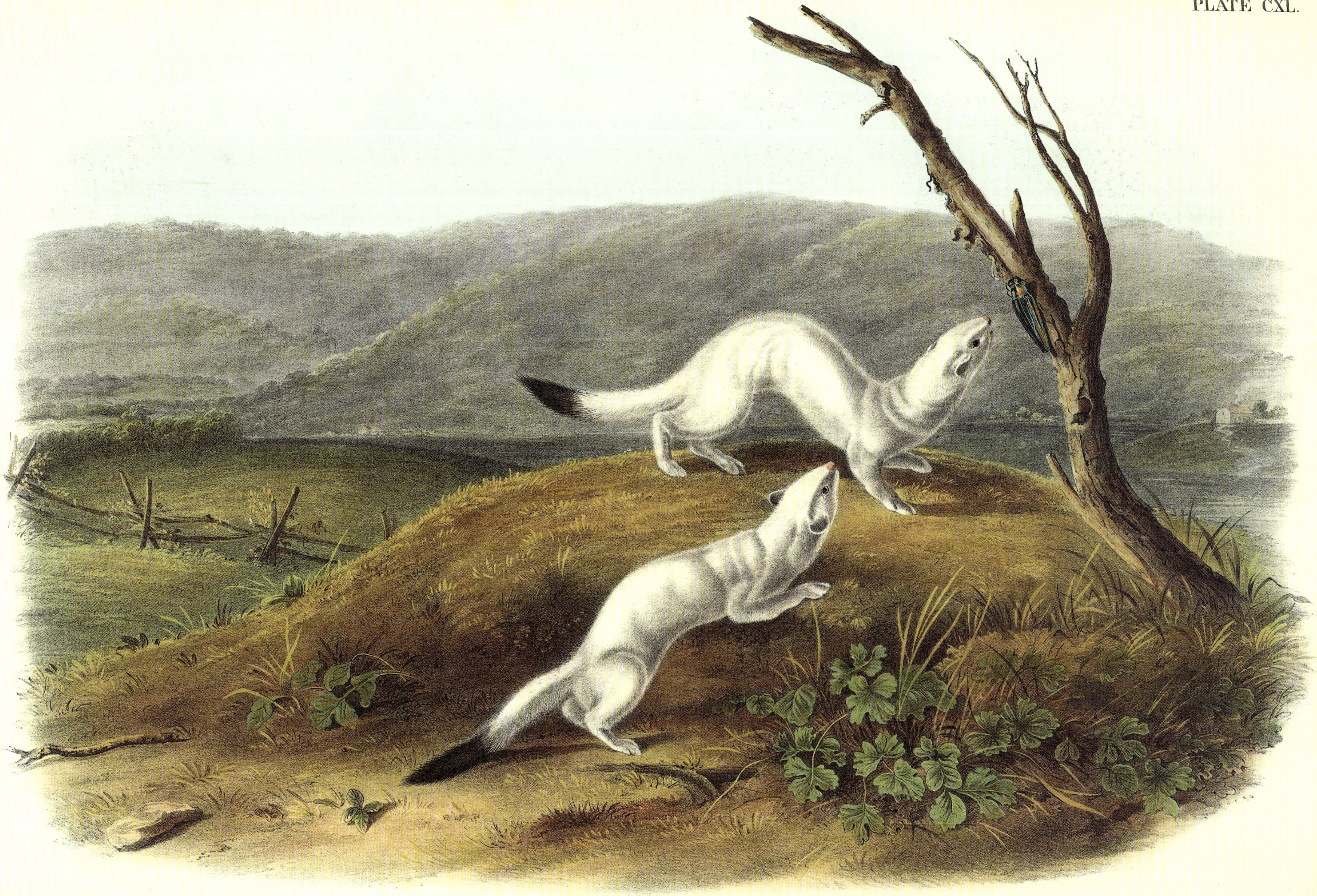

Drawn from Nature by J.W. Audubon

Lith.d Printed & Col.d by J. T. Bowen, Philad.a 1848.

PUTORIUS AGILIS, AUD & BACH.

LITTLE NIMBLE WEASEL

MALE & FEMALE.

Natural Size.

WINTER PELAGE

Nº 29.

PLATE CXLI.

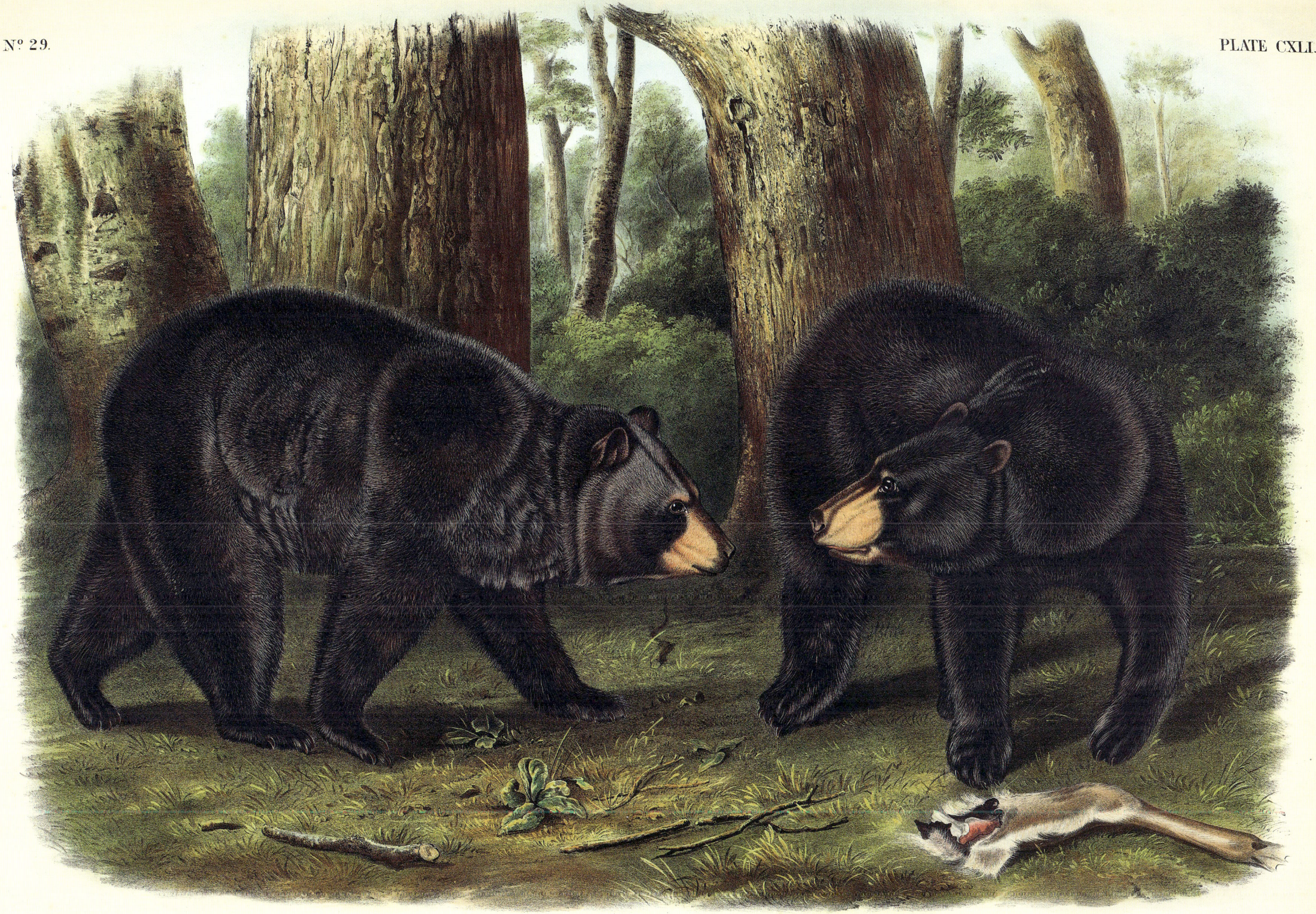

Drawn from Nature by J. W. Audubon.

URSUS AMERICANUS, PALLAS.

AMERICAN BLACK BEAR.

MALE & FEMALE.

Lith. Printed & Col. by J. T. Bowen, Philad. 1848

PLATE CXLII.

PSEUDOSTOMA BOREALIS, RICH. Males & Females.

THE CAMAS RAT.

MALE, FEMALE & YOUNG.

Natural Size.

Drawn from Nature by J. W. Audubon.

Lith Printed & Col⁴ by J.T.Bowen, Philad⁴ 1848.

PLATE CXLIII.

Fig 1.

Fig 2.

Drawn from Nature by J.W. Audubon

Fig 1. PTEROMYS SABRINUS, PENNANT. | Fig 2. PTEROMYS ALPINUS, RICH.

SEVERN RIVER FLYING SQUIRREL. | ROCKY MOUNTAIN FLYING SQUIRREL.

Lith. Printed & Col⁴ by J.T. Bowen, Philad⁴ 1848.

PLATE CXLIV.

N°.29.

Drawn from Nature by J. W. Audubon.

Lith. Printed & Col.d by J. T. Bowen, Philad.a 1848.

Fig.1. ARVICOLA TOWNSENDII, BACH.
TOWNSEND'S ARVICOLA.
MALE.
Natural Size.

Fig 2. ARVICOLA NASUTA, AUD & BACH.
SHARP-NOSED ARVICOLA.
Natural Size.

Fig.3 MUS RIPARIUS, AUD & BACH.
BANK RAT.
Natural Size.

PLATE CXLV.

Drawn from Nature by J. W. Audubon.

Lith⁴ Printed & Col⁴ by J.T. Bowen, Philad⁴ 1848

SCALOPS TOWNSENDII, BACH.
TOWNSEND'S SHREW MOLE.
MALES.
Natural Size.

N° 30.

PLATE CXLVI.

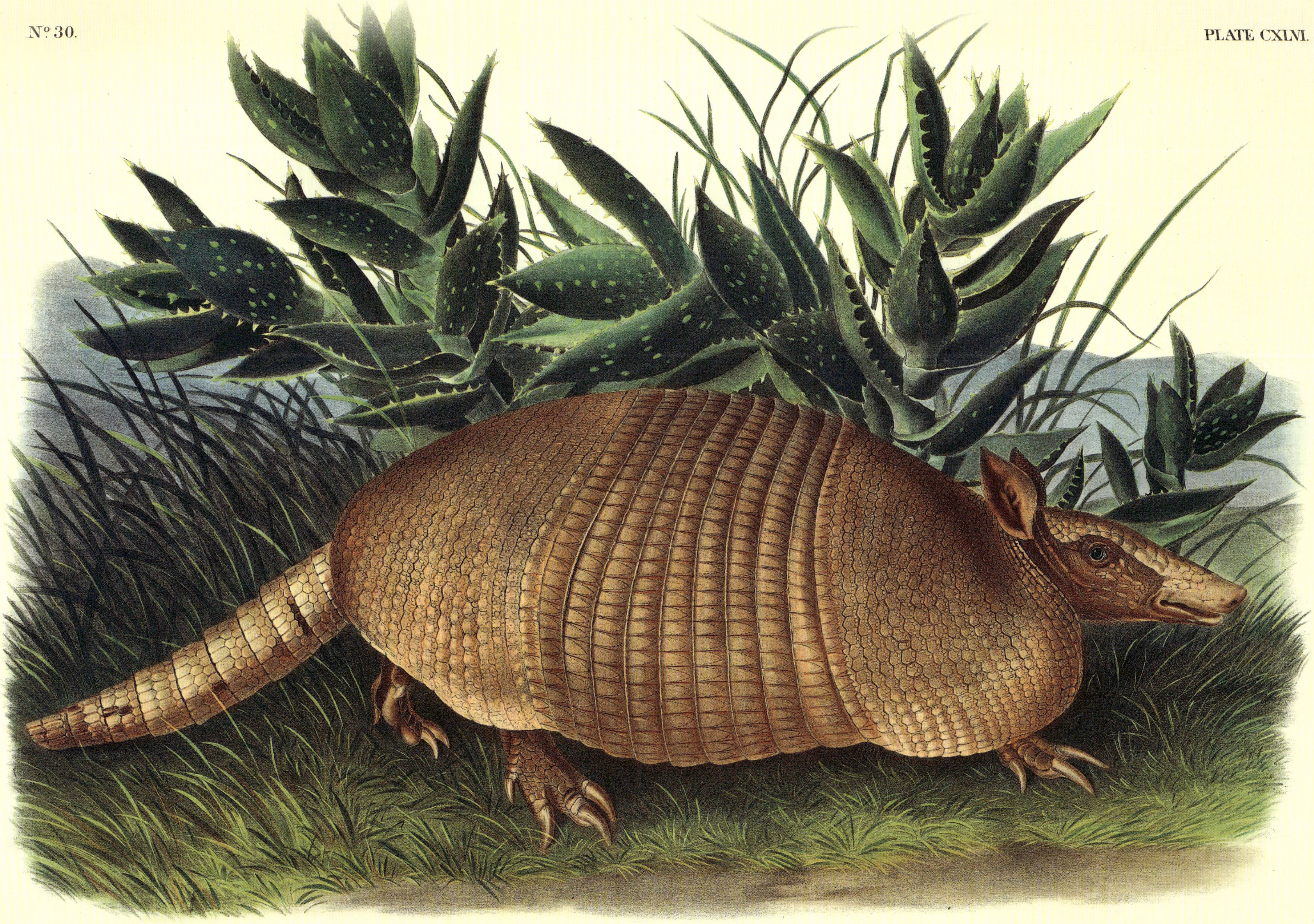

Drawn from Nature by J. W. Audubon.

Lith.d Printed & Col.d by J. T. Bowen, Philad.a 1848.

DASYPUS PEBA, DESM.

NINE-BANDED ARMADILLO.

MALE.

Natural Size.

PLATE CXLVII.

N.º 30.

Fig 3.

Fig 2.

Fig 1.

Drawn from Nature by J.W.Audubon.

Lith.Printed & Col.d by J. T. Bowen, Philad.a 1848.

Fig. 1. SPERMOPHILUS TOWNSENDII, BACH.
AMERICAN SQUIRREL.
MALE.
Natural Size.

Fig 2. ARVICOLA OREGONI. BACH.
OREGON MEADOW MOUSE.
MALE.

Fig 3. ARVICOLA TEXIANA, AUD.& BACH.
TEXAN MEADOW MOUSE.
MALE.
Natural Size.

PLATE CXLVIII.

PUTORIUS FUSCUS, AUD. & BACH.

TAWNY WEASEL.

MALE.
Natural Size.

Drawn from Nature by J.W.Audubon.

Lith. Printed & Col. by J.T.Bowen, Philad.ª 1848.

PLATE CXLIX.

Fig 1.

Fig.2.

Drawn from Nature by J.W.Audubon

Lith'd, Printed & Col'd by J. T Bowen, Phila. 1848.

Fig.1. SCIURUS FREMONTII, TOWNSEND.
FREMONT'S SQUIRREL.
Natural Size.

Fig 2. SCIURUS FULIGINOSUS, BACH.
SOOTY SQUIRREL.
Natural Size.

PLATE CL.

Nº 30.

Fig 3

Fig 2

Fig 4

Fig 1

Drawn from Nature by J.W. Audubon.

Lith.d Printed & Col.d by J.T.Bowen, Phila.ᵈ 1848.

Fig 1 PSEUDOSTOMA FLORIDANA, AUD & BACH.
SOUTHERN POUCHED RAT.
OLD MALE.
Natural Size.

Fig 2 SOREX DEKAYI, BACH.
DEKAY'S SHREW.
YOUNG MALE.
Natural Size.

Fig 3 SOREX LONGIROSTRIS, BACH.
LONG-NOSED SHREW.
MALE.
Natural Size.

Fig 4 SCALOPS ARGENTATUS, AUD. & BACH.
SILVERY SHREW MOLE.
FEMALE.
Natural Size.

Plate Information

The information below is as follows: the common and scientific species names as given by Audubon, followed by the current common and scientific species names. The numbering of the plates is as it was originally. We would like to thank Professor Kristofer Helgen for using his unrivalled expertise in American mammal taxonomy to make sure that the current common and scientific species names are completely up to date.

PLATE NUMBER
Audubon common name
Audubon species name
Current common name
Current species name

PLATE I
Common American wild cat
Lynx rufus
Bobcat
Lynx rufus

PLATE II
Maryland, marmot, woodchuck, groundhog
Arctomys monax
Woodchuck
Marmota monax

PLATE III
Townsend's rocky mountain hare
Lepus townsendii
White-tailed jackrabbit
Lepus townsendii

PLATE IV
Florida rat
Neotoma floridana
Eastern woodrat
Neotoma floridana

PLATE V
Richardson's Columbian squirrel
Sciurus richardsonii
Red Squirrel
Tamiasciurus hudsonicus richardsoni

PLATE VI
American cross fox
Canis (vulpes) fulvus
Red fox
Vulpes vulpes fulvus

PLATE VII
Carolina grey squirrel
Sciurus carolinensis
Eastern gray squirrel
Sciurus carolinensis

PLATE VIII
Chipping squirrel
Tamias lysteri
Eastern chipmunk
Tamias striatus lysteri

PLATE IX
Parry's marmot squirrel
Spermophilus parryi
Arctic ground squirrel
Urocitellus parryii

PLATE X
Common American shrew mole
Scallops aqualicus
Eastern mole
Scalopus aquaticus

PLATE XI
Northern hare
Lepus americanus
Snowshoe hare
Lepus americanus

PLATE XII
Northern hare
Lepus americanus
Snowshoe hare
Lepus americanus

PLATE XIII
Musk-rat or musquash
Fiber Zibethicus
Common muskrat
Ondatra Zibethicus

PLATE XIV
Hudson's Bay squirrel chickaree red squirrel
Sciurus hudsonius
Red squirrel
Tamiasciurus hudsonicus

PLATE XV
Oregon flying squirrel
Pteromys oregonensis
Humboldt's flying squirrel
Glaucomys oregonensis

PLATE XVI
Canada lynx
Lynx canadensis
Canadian lynx
Lynx canadensis

PLATE XVII
Cat squirrel
Sciurus cinereus
Eastern fox squirrel
Sciurus niger cinereus

PLATE XVIII
Marsh hare
Lepus palustris
Marsh rabbit
Sylvilagus palustris

PLATE XIX
Soft-haired squirrel
Sciurus mollipilosus
Douglas's squirrel
Tamiasciurus douglasii mollipilosus

PLATE XX
Townsend's ground squirrel
Tamias townsendii
Townsend's chipmunk
Tamias townsendii

PLATE XXI
Grey fox
Canis (vulpes) virginianus
Gray fox
Urocyon cinereoargenteus

PLATE XXII
Grey rabbit
Lepus sylvaticus
Eastern cottontail
Sylvilagus floridanus

PLATE XXIII
Black rat
Mus rattus
Roof rat
Rattus rattus

PLATE XXIV
Four-striped ground squirrel
Tamias quadrivittatus
Colorado chipmunk
Tamias quadrivittatus

PLATE XXV
Downy squirrel
Sciurus lanuginosus
Red squirrel
Tamiasciurus hudsonicus lanuginosus

PLATE XXVI
The wolverine
Gulo luscus
Wolverine
Gulo gulo luscus

PLATE XXVII
Long haired squirrel
Sciurus longipilis
Western gray squirrel
Sciurus griseus
NB: ID not confirmed

PLATE XXVIII
Common flying squirrel
Pteromys volucella
Southern flying squirrel
Glaucomys volans

PLATE XXIX
Rocky mountain neotoma
Neotoma drummondii
Bushy-tailed woodrat
Neotoma cinerea

PLATE XXX
Cotton rat
Arvicola hispidus
Hispid cotton rat
Sigmodon hispidus

PLATE XXXI
Collared peccary
Dycoteles torquatus
Collared peccary
Pecari tajacu

PLATE XXXII
Polar hare
Lepus glacialis
Arctic hare
Lepus arcticus

PLATE XXXIII
Mink
Putorius vison
American mink
Mustela vison

PLATE XXXIV
Black squirrel
Sciurus niger
Eastern fox squirrel
Sciurus niger

PLATE XXXV
Migratory squirrel
Sciurus migratorius
Eastern gray squirrel
Sciurus carolinensis

PLATE XXXVI
Canada porcupine
Hystrix dorsata
North American porcupine
Erethizon dorsatum

PLATE XXXVII
Swamp hare
Lepus aquaticus
Swamp rabbit
Sylvilagus aquaticus

PLATE XXXVIII
Red-bellied squirrel
Sciurus ferruginiventris
Red-bellied squirrel
Sciurus aureogaster

PLATE XXXIX
Leopard spermophile
Spermophilus tridecemlineatus
Thirteen-lined ground squirrel
Ictidomys tridecemlineatus

PLATE XL
White footed mouse
Mus leucopus
White-footed deermouse
Peromyscus leucopus

PLATE XLI
Pennant's marten or fisher
Mustela canadensis
Fisher
Pekania pennanti

PLATE XLII
Common American skunk
Mephitis americana
Striped skunk
Mephitis mephitis

PLATE XLIII
Hare squirrel
Sciurus leporinus
Western gray squirrel
Sciurus griseus

PLATE XLIV
Canada pouched rat
Pseudostoma bursarius
Plains pocket gopher
Geomys bursarius

PLATE XLV
Wilsons meadow mouse
Arvicola pennsylvanicus
Meadow vole
Microtus pennsylvanicus

PLATE XLVI
American beaver
Castor fiber americanus
American beaver
Castor canadensis

PLATE XLVII
American badger
Meles labradoria
American badger
Taxidea taxus

PLATE XLVIII
Douglass squirrel
Sciurus douglassii
Douglas's squirrel
Tamiasciurus douglasii

PLATE XLIX
Douglasse's spermophile
Spermophilus douglassii
California ground squirrel
Otospermophilus beecheyi douglasii

PLATE L
Richardson's spermophile
Spermophilus richardsonii
Richardson's ground squirrel
Urocitellus richardsonii

PLATE LI
Canada otter
Lutra canadensis
North American river otter
Lontra canadensis

PLATE LII
Swift fox
Vulpes velox
Swift fox
Vulpes velox

PLATE LIII
Texan skunk
Mephitis mesoleuca
American hog-nosed skunk
Conepatus leuconotus

PLATE LIV
Brown or Norway rat
Mus decumanus
Brown rat
Rattus norvegicus

PLATE LV
Red-tailed squirrel
Sciurus rubricaudatus
Eastern fox squirrel
Sciurus niger rufiventer

PLATE LVI
American bison or buffalo
Bos americanus
American bison
Bison bison

PLATE LVII
American bison or buffalo
Bos americanus
American bison
Bison bison

PLATE LVIII
Orange-bellied squirrel
Sciurus sub-auratus
Eastern fox squirrel
Sciurus niger subauratus

PLATE LIX
White weasel or stoat
Putorius erminea
Ermine
Mustela erminea

PLATE LX
Bridled weasel
Putorius frenata
Long-tailed weasel
Mustela frenata

PLATE LXI
Raccoon
Procyon lotor
Raccoon
Procyon lotor

PLATE LXII
American elk or wapiti deer
Cervus canadensis
Elk (Wapiti)
Cervus elaphus canadensis

PLATE LXIII
Black-tailed hare
Lepus nigricaudatus
White-sided jackrabbit
Lepus callotis

PLATE LXIV
Little American brown weasel
Mustela fusca
Long-tailed weasel
Mustela frenata noveboracensis

PLATE LXV
Little harvest mouse
Mus minimus
Eastern harvest mouse
Reithrodontomys humulis

PLATE LXVI
Virginian opossum
Didelphis virginiana
Virginia opossum
Didelphis virginiana

PLATE LXVII
Black American wolf
Canis lupus
Wolf
Canis lupus

PLATE LXVIII
Fox squirrel
Sciurus capistratus
Eastern fox squirrel
Sciurus niger

PLATE LXIX
Common star-nose mole
Condylura cristata
Star-nosed mole
Condylura cristata

PLATE LXX
Say's least shrew
Sorex parvus
North American least shrew
Cryptotis parva

PLATE LXXI
Prairie wolf
Canis latrans
Coyote
Canis latrans

PLATE LXXII
White American wolf
Canis lupus var. albus
Wolf
Canis lupus

Index

[Current common name and plate number]